TESTED
ELECTRONICS
TROUBLESHOOTING
METHODS

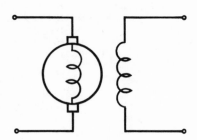

TESTED
ELECTRONICS
TROUBLESHOOTING
METHODS

Walter H. Buchsbaum

Prentice-Hall, Inc.
Englewood Cliffs, N.J.

Prentice-Hall International, Inc., *London*
Prentice-Hall of Australia, Pty. Ltd., *Sydney*
Prentice-Hall of Canada, Ltd., *Toronto*
Prentice-Hall of India Private Ltd., *New Delhi*
Prentice-Hall of Japan, Inc., *Tokyo*

©1974 by
Prentice-Hall, Inc.
Englewood Cliffs, N.J.

Second Printing . . . . . May, 1975

Library of Congress Cataloging in Publication Data

Buchsbaum, Walter H
  Tested electronics troubleshooting methods.

  Includes bibliographical references.
  1. Electronic apparatus and appliances--Maintenance
and repair. I. Title.
TK7870.B8   621.381   74-10719
ISBN 0-13-906958-5

Printed in the United States of America

About This Book

The host of different machines, devices, and equipment that make modern life so easy and efficient are all vulnerable to breakdown, and when that occurs, no person is more important and more urgently needed than a competent troubleshooter. All of us know of electronics technicians who spend expensive hours in vain attempts to troubleshoot a relatively simple defect. We have all had occasion to lose confidence and to despair of ever getting a particular thing repaired. But then there is the genius who can tell just what is wrong in a few minutes and repair a defect—efficiently and at reasonable cost. This type of individual, unfortunately, is extremely rare.

The question is frequently asked, what makes the difference between the competent, efficient troubleshooter and the average repairman who is able to replace obviously defective parts but who is totally helpless when a more subtle defect occurs. Is the successful technician specially gifted? Does he have an inborn knack, or is it something that he has been lucky to acquire? Can anyone learn this knack?

The answer is quite simple. Yes, anyone who can learn to understand the principles of the electrical appliance or electronic device can also learn to troubleshoot these things quickly and efficiently. This book will help the technician who has a fundamental knowledge of electronics to acquire the knack and the skill to troubleshoot any kind of electronic equipment efficiently, successfully, and profitably.

We have observed quite a few successful technicians and have compared their methods of operation with the run-of-the-mill, screwdriver mechanic variety, and we have found what makes the difference. In many instances, the apparently "inborn" knack is nothing more than a method or technique the technician has acquired by learning from his own experience. In many other cases, the successful technician has acquired his troubleshooting technique by observing others, and, in a very large number of cases, the "knack" is the result of effective training, either in the military services or in a very good vocational school. You can acquire the knack to troubleshoot electronic equipment skillfully by studying this book and by using it as a reference when you encounter unexpected difficulties.

To get the most out of this book, you should know how to read electronic circuit diagrams and be familiar with the fundamentals of electricity, such as Ohm's law, and some of the fundamental components of electronics, such as vacuum tubes and semiconductors. You do not need mathematics, modern network theory, or any of the other skills involved in original equipment design. You should be able to read manufacturer's instructions and use standard electronic test equipment, such as the oscilloscope and volt-ohmmeter. With these fundamentals as a background, this book will help you acquire successful troubleshooting techniques and show you where each technique is most useful and how each is best applied.

A few basic troubleshooting techniques form the foundation on which all successful troubleshooting work rests. Once these techniques are understood, this book will show you how to get the most from your test equipment. You will learn what the key techniques are in troubleshooting transistor circuits, integrated circuits, and electron tube circuits. Because intermittent defects are usually very hard to find, a special chapter of this book is devoted to techniques for dealing with this type of trouble. Another chapter is aimed at those defects that are due to some kind of interference and are often mistaken for component failures. The last three chapters cover the special aspects of troubleshooting television and hi-fi equipment, digital devices and computers, and industrial controls and instruments.

Once you have completed the first five chapters, you will be able to troubleshoot any electronic equipment with much greater efficiency and a much better chance of success. The second half of the book gives you the information needed for specific types of defects and specific types of equipment and is invaluable as a reference later on when you encounter a particular problem.

Whether you are an experienced technician interested in troubleshooting more efficiently and profitably, or whether you are a beginner, this book will help you acquire that envied, much sought-after reputation of the skilled electronics troubleshooter who has the "knack" to do the job well and quickly.

Walter H. Buchsbaum

Acknowledgments

Although only a single author is credited with this book, it is really the product of the cooperative efforts of many different people. I want to thank all of the professional troubleshooters who have contributed from their store of experience, and the many electronics manufacturers who supplied technical material concerning their products. Where specific illustrations are obtained from a manufacturer, a credit line shows the source. I want to thank Mrs. Inge Seymour for the beautifully typed manuscript.

Finally, this book would not have been possible without the forebearance and encouragement of my wife, Ann, who also arranged the index.

Contents

TROUBLESHOOTING TRANSISTOR CIRCUITS *(cont.)*

4   TROUBLESHOOTING INTEGRATED CIRCUITS 78

5   TROUBLESHOOTING ELECTRON TUBE CIRCUITS.............105

6   FINDING INTERMITTENT DEFECTS 127

TROUBLESHOOTING DIGITAL DEVICES AND COMPUTERS *(cont.)*

10   TROUBLESHOOTING INDUSTRIAL CONTROLS AND INSTRUMENTS..205

Basic Troubleshooting Techniques

1.1 HOW TECHNIQUES GET RESULTS AND MAKE PROFITS

If you were asked to find an error on this page, how would you go about it? Would you start by carefully reading every word on this page? Would you glance over the page quickly to see if it is a printing error, such as a line or a letter out of the margin? Would you look for an error in spelling, an error in grammar, or an error in technical content?

One method may be to read the entire page, word by word, line by line, and look up, in a dictionary, every word that might be misspelled. This method would probably pick up any typographical or spelling errors. Another method is to skim over the page and get the meaning of each sentence so that you can decide whether there is a contradiction in meaning or some technical inaccuracy. A third method would be to ask what kind of an error it is. Once you know it is an error in spelling, or in grammar, etc., you can then use the best method for locating that particular type of error.

Troubleshooting techniques for electronic equipment are really just an effort to be most efficient in locating a particular defect, or error, in the electronic equipment. It is possible to find defects by a meticulous search and analysis of the performance of the equipment. It is also often possible to find defects by testing every component. Still another method concerns itself with the input and output of each functional block. Which method to use will depend largely on what type of defect we are trying to locate.

In troubleshooting electronic equipment the old saw that "time is money" is absolutely and always true, since the cost of troubleshoot-

13

ing depends entirely on the time spent in locating the trouble. The cost of replacement parts is the same whether it takes us ten minutes or ten hours to find the defective component. The faster you can do the troubleshooting, the more profit you can make.

As we have seen in the example of the supposed error on the previous page, the method used to find the trouble usually determines how long it will take. It would be foolish, for example, to take elaborate voltage and resistance measurements on a nonoperating TV receiver if the owner has indicated that smoke came out of the set and he therefore shut it off. Obviously, the first thing to do is to look for the component that produced the smoke. This immediately isolates the defect to a particular section. On the other hand, hours of visual inspection will yield nothing if the defect is an intermittent fading in the colors of a picture or the amplitude of the sound.

1.2 HOW HELPFUL ARE MANUFACTURERS' DATA?

If you are driving around your hometown, a road map is quite useless to you since you know the way already. When you are driving in strange territory, however, a road map is of great help. Of course, the road map should be printed in English and should be accurate enough to show you all of the streets, with their names, through which you must drive. Manufacturers' data are essentially the same as road maps.

If you are thoroughly familiar with a particular model of the electronic equipment you are trying to troubleshoot, you can safely disregard the manufacturers' circuit diagram. If you know what all of the voltages, resistances, signals, and and other pertinent information should be, you also don't need any of the manufacturers' data. Unfortunately, many of the data packages supplied by manufacturers are often not as helpful to the troubleshooter as they should be. Occasionally, instead of giving the actual values of resistors, capacitors, transistors, etc., the manufacturer merely prints his own part number, which does not really help the troubleshooter. There is often no indication of waveforms, amplitudes, timing sequences, and so on.

However, even inadequate road maps are often better than none because they allow you to orient yourself, at least partially, with regard to the actual situation. Similarly, manufacturers' data are always of some help, even if they only help you draw a good circuit diagram. Of course, the more you understand the operation of the equipment you are troubleshooting, the better you can utilize the principles of electronics and the better your troubleshooting technique will be. The manufacturers' data will likewise become more useful. Just as the American Automobile Association provides maps specifically for the driver in strange territory, so do a number of technical service companies furnish troubleshooting data, circuit diagrams, etc., which are usually more detailed and more useful than the "free" data supplied by the manufacturer. Many experienced electronic troubleshooters therefore rely more on the material supplied by these technical services, even though they cost a few dollars, than the "free" manufacturers' data.

1.3   FITTING THE TECHNIQUE TO THE PROBLEM

As we have illustrated before with the supposed error on the page, the type of trouble really determines which technique is the most efficient one to solve the particular problem. Just as a mechanic learns early to use the right tool, so will the electronics troubleshooter find that it pays to use the right technique. The problem very often, however, is to decide which technique will be best to troubleshoot the defect, particularly if we are not sure just how the trouble is caused and what causes it. It is obviously impossible to predict exactly all of the defects that can occur. What we can do, however, is to classify the most frequent troubles in electronic equipment and to advise the reader which technique works best with a particular class of defects.

In difficult cases it may be necessary to use several techniques. First, the symptom-function technique may be used to isolate the defect to a particular portion of a complex piece of electronic equipment. Once the trouble spot is isolated, the signal-tracing technique may be used to find a particular stage, amplifier, logic gate, or whatever is the actual cause of the malfunction. A third technique, measuring voltage and resistance, may be required to isolate a particular component. As a last resort, the substitution technique may still be required to make sure that, among several components, the real culprit is found. There are also defects that resist the application of these four basic tech-

15

niques. At the end of this chapter we will discuss which measures the troubleshooter can take for these troubles.

Before we can begin to decide which technique to use for a particular troubleshooting task, we should try to find out as much as we can about the trouble itself. In the "error on this page" example, if you could have asked what type of error it was, it certainly would have helped you find it much faster. To help you troubleshoot a particular electronic equipment it is usually possible to ask the person reporting the trouble a number of questions that can be of great help in the actual troubleshooting process and in the selection of the best technique.

Table 1-1 shows a typical list of ten questions. The first question is intended to elicit a specific description of the trouble. When the owner of a hi-fi set says it doesn't work right, this provides very little information. If we ask him why he thinks it does not work right or which particular feature does not work right, we might learn that one of the two channels of the stereo system seems to be much weaker than the other. Knowing which channel it is immediately narrows the problem down to one of measuring audio gain in that channel. If the owner of a TV receiver reports that nothing happens when the set is turned on, and explains that neither sound nor picture nor even a raster is obtained, we will immediately suspect the power supply, a fuse, or circuit breaker malfunction.

The second question is aimed at further focusing on the defect. In the case of the stereo system, we may ask the owner whether he has tried adjusting the volume control, loudness control, tone control, or balance. In the case of the TV set, we may ask whether some other electrical appliance plugged into the same socket works well, or whether it is possible to see any light in the filaments of the picture tube or any other tube.

The third question is intended to tell us whether the defect is intermittent, or whether it depends on some external effect or whether it is a clear-cut failure. If it is intermittent, the fourth question tries to establish under what conditions the defect occurs. Sometimes defects depend on shock or vibration, temperature, or some apparently random effect. Question five helps us to know if the defect has only been apparent since a particular equipment has been dropped, vibrated by transportation in an automobile, has overheated, or has been similarly abused.

1. What is *really* wrong?
2. How is this defect apparent?
3. Is it always this way?
4. If intermittent, under what conditions?
5. Was there any abuse? (Vibration, shock, heat, etc.)
6. Did the defect occur suddenly or gradually?
7. Did the defect occur during equipment operation?
8. Does the defect appear to affect other functions?
9. Any additional details?
10. Has anyone tried to fix it?

TABLE 1-1

The sixth question is intended to tell us if the defect may be due to aging or whether it is a sudden failure. Similarly, the seventh question might help us determine if the defect is in any way associated with the surges that frequently occur during equipment turn-on or turn-off.

Sometimes the defect of a particular function also affects others, and this may indicate the exact nature of the defect quickly. In a TV receiver, for example, we may be told that the picture becomes fuzzy and dim as the primary defect. We may also learn that, at the same time, the sound became weak and the volume control had to be turned to maximum. This would indicate a power supply failure rather than a defect in the video itself.

The ninth question is intended to bring out any additional details that could help us locate the defect. TV picture distortion, coinciding with the operation of a vacuum cleaner, for example, falls into this category. Finally, we know that any time an unskilled person tries to make any sort of repair, this is likely to complicate the task of the electronics troubleshooter. The tenth question will solve this.

With the knowledge acquired as a result of questions like those listed in Table 1-1, you can usually make a pretty good prediction as to where the trouble is and which of the basic four techniques is most likely to yield quick results. Much depends, of course, on the type of equipment you are dealing with. For this reason, the detailed practical information contained in chapters 8, 9, and 10 will be invaluable at this point. First, however, the more general aspects of when and how to use each of the basic four techniques will be described.

1.4 WHEN AND HOW TO USE THE SYMPTOM-FUNCTION TECHNIQUE

Of the four basic techniques described in this book, the symptom-function technique is probably the most powerful and will be used whenever possible. It requires practically no test equipment and depends mostly on the small but highly efficient computer we carry inside our heads.

We already use the symptom-function technique in many daily tasks. If you drive along in your car and the left front is suddenly lower and makes a funny, flapping noise, you know at once that you have a flat left front tire. You know very well that this kind of symptom could not possibly come from a malfunction of the engine, the drive shaft, the differential, or any other part of the car. You have automatically associated the symptom with the function.

When you turn the switch on a table lamp and it fails to light, you will immediately try switching again, checking to make sure that the AC power cord is properly connected and then change the bulb. If a new bulb fails to light, and if another appliance plugged into the same socket works properly, only then will you suspect a problem in the switch and possibly the wiring itself. You have used the symptom-function technique by associating the observed symptom, the fact that the lamp does not light, with a function of the different elements causing it to light—the AC power coming in through the cord, the bulb itself, and, lastly, the switch and the wiring.

In troubleshooting electronic equipment with the symptom-function technique, the relationship between the symptoms, the functions, and the defect are usually not as simple to understand and as easy to remember as in the examples above. By understanding the principles of a particular piece of electronic equipment and by visualizing the various functions and how they are performed, however, it is often possible to find the trouble by reason alone, without test equipment and without taking measurements. The two block diagrams in Figure 1-1 show typical fundamentals of operation of many kinds of electronic equipment, ranging from hi-fi sets or television receivers, to complex digital circuits. Figure 1-1a illustrates the case where there are a number of different inputs, all going to a single output. It is clear that we can isolate the trouble effectively if we know which of the inputs does not produce an output. Similarly, Figure 1-1b shows the case of

18

a single input and three different outputs. Here, too, it is easy to localize the trouble by observing which outputs work and which do not. The third possible arrangement is where there are a number of inputs as well as a number of outputs and where the relationships between them are somewhat more complex.

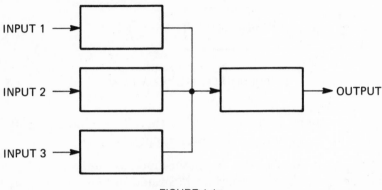

FIGURE 1-1a
MULTIPLE INPUT—SINGLE OUTPUT

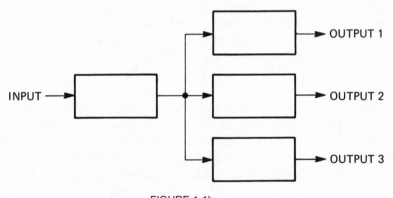

FIGURE 1-1b
SINGLE INPUT—MULTIPLE OUTPUT

First, consider the case of a typical FM-phono hi-fi system as shown in the block diagram in Figure 1-2. Clearly here we have two different signal inputs—the phono pickup and the FM tuner. The output is the output amplifier and the speaker it drives. If this were a stereo system, there would be two preamps, two output amplifiers, and two speakers, and this would clearly complicate the problem. If we look at

19

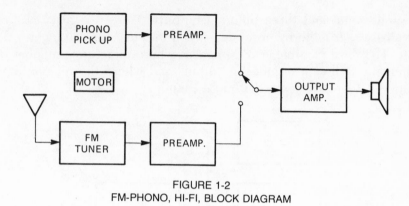

FIGURE 1-2
FM-PHONO, HI-FI, BLOCK DIAGRAM

the block diagram carefully, we realize that there is a third input that affects all of the outputs—the power supply.

A typical application of the symptom-function method would show the basic troubleshooting chart as illustrated in Figure 1-3. Only the most probable types of failures are listed in this table, but others, much less likely, are possible of course. In the case of the first item, for example, the motor may be perfectly good, but the turntable itself may be stuck or the drive mechanism may be slipping. In the second and third instances, the phono-FM switch may be defective. In the fourth case, the antenna lead may be broken, or there may be another such problem.

An example of a single input and multiple outputs is shown in Figure 1-4, the block diagram of color TV receiver. The single input is the TV antenna, which passes the signal through the tuner, the IF amplifier, and detecting stages. There the signal splits up. If the picture is satisfactory but the sound is missing, clearly the defect is in the audio or speaker section. If the audio is fine and only a black-and-white picture appears, the defect is most likely in the chroma section. Similarly, the defect can be isolated further to the sync, horizontal sweep, vertical sweep, high voltage, and power supply sections. A question arises now because a defect in the horizontal sweep section will also cause loss of high voltage and this will cause total lack of picture. It is thus possible to mistake the symptom of "no picture" as a defect of the video detector function. Clearly at this point it would be necessary to switch from the symptom-function technique to one of the other techniques.

Symptom	Function Defective
Turntable doesn't turn	
FM sound okay	Motor
No phono sound	Phono pickup or
FM sound okay	Phono preamp
No phono sound	Power supply or
No FM sound	Output amp or
Turntable turns	Speaker
No FM sound	FM Tuner or
Phono sound okay	Preamp

FIGURE 1-3

If the television set uses vacuum tubes, we would probably switch to the substitution technique, described in more detail in section 1.7 below, and substitute a horizontal output amplifier tube to see whether this has caused the lack of high voltage, horizontal sweep, and picture all together. If the defect were a single horizontal line, then the symptom-function technique would immediately suggest a failure in the vertical sweep section, and, again, by substituting another vertical sweep tube we should be able to effectively troubleshoot this defect.

The power supply shown in the block diagram in Figure 1-4 provides different voltages for different sections of the color TV receiver. A failure in the transformer primary would, of course, disable all sections and the symptom would be readily recognized. None of the tubes, including the picture tube, would be illuminated and we would suspect a failure in the primary portion of the power supply. If, however, the power supply portion fails, which supplies power to the horizontal sweep section, then it will be more difficult to find the defect. Substitution of a new tube in the horizontal sweep section would not help. In this case we would have to switch to the voltage-resistance technique described in more detail in section 1.6 below, and measure the voltage going to the horizontal sweep section.

The examples described above show that the symptom-function technique is usually used as the first approach in isolating the trouble.

21

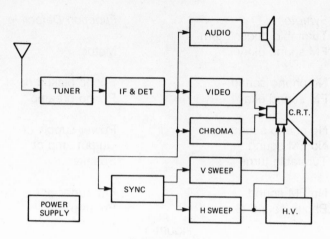

FIGURE 1-4
COLOR TV RECEIVER BLOCK DIAGRAM

Sometimes it is possible to determine the defect by reasoning alone, but, most frequently, one of the other techniques must be used to hunt down the defective component. The real efficiency of the symptom-function technique lies in its ability to localize troubles. This means that instead of having to make many different measurements, or having to substitute many different parts or tracing signals through many different circuits, the troubleshooting can be localized to a particular tube, transistor, or integrated circuit stage.

The symptom-function technique should always be used first. It requires a knowledge of the functions performed by the major sections of the equipment you are troubleshooting, and it requires us to associate particular symptoms with the function. If you are not familiar with the functions, a block diagram or a circuit diagram should be consulted and each of the symptoms the set owner reports should be confirmed by your own observation.

TO SUM UP, USE THE SYMPTOM-FUNCTION TECHNIQUE FIRST. UNDERSTAND THE FUNCTIONS PERFORMED BY THE EQUIPMENT. OBSERVE THE SYMPTOMS AND ASSOCIATE FUNCTIONS TO LOCALIZE THE DEFECT.

1.5 WHEN AND HOW TO USE THE SIGNAL-TRACING TECHNIQUE

If the symptom-function troubleshooting method has localized the defect and preliminary tests do not reveal the source of the defect, it is often necessary to trace the trouble by means of the signal-tracing method. The detailed descriptions in later chapters cover the many variations of signal tracing in different types of equipment, but the principles explained below apply to all signal tracing.

Figure 1-5 illustrates the principle of signal tracing in a basic amplifier. Whether it amplifies DC, audio, video, IF, etc., a signal generator with an internal source resistance R_G provides the input signal to the amplifier. The amplitude of the input signal is measured by V_I as it appears across the input impedance R_I. The output of the amplifier is measured by V_O as it appears at the load resistor R_L.

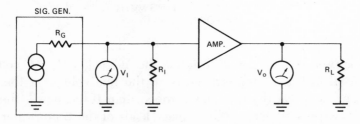

FIGURE 1-5
SIGNAL TRACING A SIMPLE AMPLIFIER

By comparing the readings of V_I and V_O, we can determine the amplification or gain of the amplifier. If the output amplitude of the signal generator is varied, we can see whether the amplifier is linear over a range of input signals. By varying the load impedance RL, we can see whether the gain of the amplifier is linear with variations of the output load. By varying the frequency of the signal generator, we can determine the frequency response of the amplifier. With this relatively simple arrangement, the important characteristics of the amplifier can be measured because we are tracing the signal, in amplitude and possibly in frequency, from the input to the output of the amplifier.

In many types of electronic equipment it is not necessary to provide a source of signal, particularly if a signal is known to be present or if its presence can easily be determined. This type of signal-tracing method is generally called "passive" and a typical example of it is shown in Figure 1-6. In this simplified circuit of a power supply we can consider the 60 Hz power line frequency as the signal to be traced.

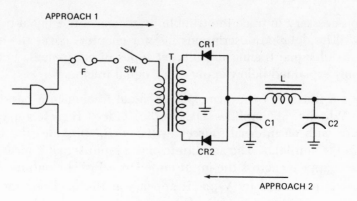

FIGURE 1-6
PASSIVE SIGNAL-TRACING METHOD OF
A POWER SUPPLY

Under approach 1, the test will start at the AC socket. An AC voltmeter is used in this example to measure the line voltage at the wall
outlet and, again at the fuse and at the switch. If the signal, the 110
volt AC, is present across the primary leads of the transformer, we
know that the plug, linecord, fuse, and switch are good. At the secondary of the transformer T, the AC signal can be measured from either
side of the secondary to ground. If the correct voltage is measured at
these two points, we know that the transformer itself is good. To
continue with approach 1, we must now switch to the DC scale of the
meter and measure the voltage across C_1 and finally across C_2. If
either C_1 or C_2 are shorted, there will be no DC voltage. If the choke
L is open, there will be voltage at C_1, but none at C_2. If C_1 or C_2 are
open-circuited, or if either of the rectifiers, CR_1 and CR_2, are open
and short-circuited, the DC voltage will not measure the correct
value. In these cases, a resistance measurement would be required to
positively identify the defective component.

Approach 1 has been to start out at the source of the signal, at the AC
plug. Approach 2 takes the opposite direction and starts with the DC
measurement across C_2, followed by a DC measurement across C_1,
and so on. In the example shown in Figure 1-6, it does not appear to
matter very much whether approach 1 or approach 2 is taken, since
the test instrument, the voltmeter, can handle either side of the
signal equally well. We will see in the following examples that the
approach we choose often depends on the type of test equipment
available, or on the symptoms observed.

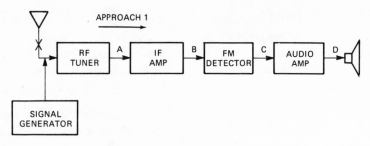

FIGURE 1-7
ACTIVE SIGNAL-TRACING METHOD OF FM RADIO

A typical example where the symptom-function approach must be replaced by the signal-tracing method is shown in Figure 1-7. The FM radio shown in the block diagram does not work, although the power supply has been checked and the proper voltages were applied. Clearly, the defect lies somewhere between the antenna and the loudspeaker.

For passive signal tracing, the presence of a "normal" signal is assumed, but since an antenna and tuning for stations is involved, we shall substitute a signal generator for the "normal" signal and use the loudspeaker as indicator of the signal. This technique is called "active" signal tracing; in some books the term "signal injection" method is used.

If approach 1 is taken, a signal generator is connected to the input of the RF tuner, with the antenna disconnected, and both signal generator and tuner are tuned to the same frequency. If nothing is heard in the loudspeaker, the signal generator is moved to point A in Figure 1-7. If the tone from the signal generator is now heard, the defect is clearly in the RF tuner. If nothing is heard, the signal generator lead is moved to the output of the IF amplifier point B, and further to point C. When moving the signal generator output from the input of the RF tuner to the input of the IF amplifier, the signal generator frequency must be changed from the FM broadcast band to the 10.7 MHz IF frequency, standard for FM radios. At point B, the amplitude of the signal generator output must be increased in order to compensate for the gain normally provided by the IF amplifier. At point C the normal signal is an audio signal, and the signal generator must then be switched or else a separate audio signal generator must be used. At point D the signal generator would have to furnish an audio signal sufficiently large to drive the speaker, something few signal generators are able to do. Fortunately, a loudspeaker can be tested by checking the voltage on the driver amplifier and momentarily reducing it by using a resistor of appropriate value between that voltage and ground. This should result in a click in the speaker.

As shown in Figure 1-8, the active signal-tracing method of our FM radio can also be performed by approach 2, starting at the speaker and gradually working toward the input of the RF tuner. In deciding whether to use approach 1 or approach 2, a few preliminary tests can be made. If, for example, we briefly short the input to the audio amplifier to ground with a screwdriver or clip lead, we should hear a click in the speaker if the audio amplifier and speaker are operating. If nothing is heard, clearly approach 2 will yield the quickest results since the defect must be either in the speaker or audio amplifier. If a click is heard, we can either use approach 2, starting at point C, or else we can decide to start with approach 1, both with equal probability of fast success.

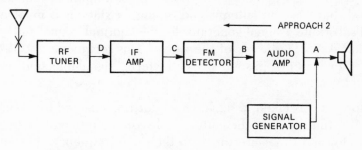

FIGURE 1-8
ACTIVE SIGNAL-TRACING METHOD OF FM RADIO

The simple examples above illustrate the principles of the signal-tracing technique. Most actual applications involve more complex equipment, such as the color TV receiver shown in block diagram form in Figure 1-4. Assume, for a moment, that the symptom-function method has shown that the most probable area of defect is in the chroma section of this color TV receiver. If a typical color bar generator is available as test equipment, it can be connected to the input of the tuner and an oscilloscope would then be required to check for the presence of the 3.58 MHz color subcarrier at the input to the chroma section. This color subcarrier would then be traced by oscilloscope through each stage of the chroma section until the defective stage or component is located.

If a color bar generator is not available, it is possible to use a signal from a station known to transmit a color telecast and signal-trace the chroma section with the oscilloscope in the same manner. For this type of signal tracing, it is necessary to know what the waveforms and amplitudes should be at each of the individual stages in the chroma section. We can then use the oscilloscope directly to observe and measure these signals. Chapter 8 deals with signal tracing in TV and hi-fi equipment and contains a more detailed description.

TO SUM UP, THE SIGNAL-TRACING METHOD REQUIRES THAT THE INPUT SIGNAL TO A SUSPECTED STAGE BE KNOWN, AND THAT WE CAN MEASURE THE OUTPUT FAIRLY ACCURATELY. SIGNAL TRACING ALWAYS REQUIRES AT LEAST ONE PIECE OF TEST EQUIPMENT AND USUALLY TWO.

1.6 VOLTAGE-RESISTANCE TROUBLESHOOTING METHOD

We have seen in the simple example in Figure 1-6, signal tracing of a power supply, that a point can be reached when one or several components are suspected and the only way to be sure which one is defective is to measure it. While test instruments are available that measure capacitance as well as inductance, the most frequently used measurement for testing suspected networks or components is that of voltage and resistance. Voltage measurements require that the equipment be "on" while resistance measurements usually require that power be turned off.

Most manufacturers' circuit diagrams and data sheets indicate the voltages required for proper operation at certain fixed test points. By measuring these voltages it is usually possible to localize the defect to a network or a particular component. In Figure 1-6, for example, the manufacturer's data may state that the voltage at C_1 is +435 volts while the voltage at C_2 is +420 volts. This immediately tells us that 15 volts are dropped across the choke L. If we know either the DC resistance of this choke or the rated current of the supply, we can determine, if these voltages are either high or low, what has happened to the load, which is not shown in Figure 1-6. A partial short circuit in the load of this power supply would reduce the voltages

27

both of C_1 and C_2 and, by measuring the difference between these two, we would know how much more current is being drawn. All this can be determined with only two DC voltage measurements.

The real value of voltage measurements will be apparent in chapters 3 and 5 when transistor and electron tube circuits are analyzed by means of the voltage-resistance method. When the voltage at the collector of a transistor is higher than specified by the manufacturer's data, this indicates that the transistor is not drawing current, most likely because of insufficient forward bias on the base. Similarly, in a vacuum tube circuit a high plate voltage indicates that the tube is cut off, most likely due to excessive negative bias on the control grid. When the voltage on the transistor collector is very low, this may indicate either excessive bias or possibly a shorted transistor. Similarly, in an electron tube low plate voltage may indicate insufficient control grid bias or even a gassy tube.

Resistance measurements are probably the single most powerful method of testing most electronic components. Simple resistance measurements can be used to establish continuity in wiring, the continuity and approximately correct values of transformers, inductors, chokes, etc., as well as the approximate value of the larger capacitor sizes. The vast majority of resistors used in electronic equipment are of the carbon composition type and they tend to change in value with aging and with the application of heat. While it is often possible to measure the resistance of a resistor or other component in the circuit, we must make sure, by checking the circuit diagram, that parallel impedances do not give us an erroneous measurement. When a resistor is either shorted or open, of course, it is relatively simple to determine this.

To check capacitors, down to 0.01 MFD, the capacitor should first be fully discharged, possibly by connecting a clip lead across it, and then the ohmmeter is connected across this capacitor with its lowest resistance range. As the voltage applied charges the capacitor, the meter needle will go from a low resistance reading to a high one. By comparing the time it takes, and the low and high ohmmeter readings with known good capacitors of the same size, you will soon get enough practice to judge whether a capacitor is of the right value. When large capacitors, 10 MFD or over, are tested, the second or third highest ohmmeter scale should be used.

The voltage-resistance technique is almost always used either after the symptom-function technique points to a particular component or network as being the source of the defect, or when a signal-tracing technique has isolated a defect in this manner.

TO SUM UP: THE VOLTAGE-RESISTANCE TECHNIQUE IS USED TO PINPOINT A DEFECTIVE NETWORK OR COMPONENT AND USUALLY REQUIRES MANUFACTURER'S DATA FOR VOLTAGES AND COMPONENT VALUES.

1.7 SUBSTITUTION METHOD

On equipment using electron tubes the easiest troubleshooting method appears to be the substitution of tubes. Even the unskilled layman is often tempted to pull out all of the tubes of a TV set, take it to one of those terrible tube testers located in drugstores and, based on the very crude tests, replace a defective tube. For the professional electronics troubleshooter, the substitution method means much more than simply plugging in new tubes. In the case of electron tubes, the mere suspicion of a defect due to the tube itself is good enough reason to substitute a known good tube because the substitution is so simple. In the case of transistors, ICs, and other components, however, the substitution usually requires soldering and this method is therefore used only after the voltage and resistance measurement technique has clearly indicated the probability of a component defect.

In a typical example, the signal-tracing method may indicate that while a certain level of audio signal is present at the output of one stage, nothing, or a much lower level, is measured at the output of the following stage. Before we start to substitute transistors, diodes, resistors, and capacitors in the stage that is obviously defective, we would first make some voltage measurements to make sure that the proper collector voltage, base bias, etc., are present. These voltage measurements, and subsequent tests of the associated resistors and capacitors, will indicate whether one of those components is bad or whether the transistor itself may be at fault. Even when the defect has been traced directly to the transistor, it is often very easy to perform some rough checks on the transistor before unsoldering it completely and soldering a new one in place. In chapter 3 we shall show how circuit parameters of transistors can be measured with a volt-ohmmeter. In any event, substituting a new transistor for one believed to be defective requires careful matching of transistor types and connections.

Substitution of components is usually the final step in the trouble-shooting process and, just as the steps before it, a clearly established, well-proven method should be used. The two essential steps in the substitution method are to use the correct replacement part and to connect that part correctly into the circuit. While this may appear obvious, in practice we are often tempted to use as a substitute something that is almost the same as the original part, usually because there is no exact replacement available.

In many cases, this works out fine. If you do not have a 680-ohm, ½-watt resistor available, but if you have a 680-ohm, 1-watt resistor, there seems to be no objection to substitute it for the defective one. If you need a .05, 100-volt capacitor, but have only a .05, 400-volt capacitor in your spare parts supply, this substitution will also work well. If, however, you want to use a 600-ohm, ½-watt, 10% resistor to substitute for a 600-ohm, ½-watt, 5% resistor, you may be successful or you may not, depending on the true value of the replacement resistor. Similarly, replacing a 100 MFD filter capacitor with a 150 MFD capacitor of the same voltage rating may work out fine but, if the time constant of the filter is critical, the additional capacity may present a problem.

In replacing transistors and diodes, exact equivalents are often not available and the semiconductor handbook you may be using often shows equivalent transistor and diode types. To avoid complicating the troubleshooting procedure, it is always good practice to actually look up the full characteristics of the original semiconductor and those of the recommended equivalent substitute. Many times it turns out that although the electrical characteristics are the same at 25° C, they may be quite different at either the higher or lower temperature end of their operating range. If you know the temperature range over which the particular semiconductor device must operate in a particular equipment, you can decide whether the replacement is sufficiently close to be used as substitute.

When it comes to replacing special types of transformers, coils, deflection yokes, and other special parts, it is particularly important to get the exact replacement. Very often a device that its manufacturer claims to be equivalent to another will operate satisfactorily for a while but, because its temperature characteristics or its power-handling ability are not the same as the original component, it may deteriorate faster or cause some other type of defect. This then makes troubleshooting the equipment the second time so much harder.

To solder a replacement resistor between the two points where the defective one was is usually no problem. Connecting a small diode as replacement may be more difficult, because the exact polarity must be observed, and the markings on the small glass case are often indistinct and can be misinterpreted. Unless the polarity of all electrolytic and tantalytic capacitors is carefully observed, the replacement unit can cause more trouble than the defective unit did originally. Similarly, in replacing transistors, integrated circuits, transformers, and most other components, it is absolutely essential to make sure that the connections are correct. Many replacement parts, especially transformers and other iron-core components, use a different color code than the original component. Sometimes the terminal pin numbers are changed. Some replacement parts often have additional terminals to make for greater universality in the replacement market. Be sure to determine from the manufacturer's data which leads, which terminals are the correct ones for the particular equipment in which you use a replacement part. Replacement transformers often have open windings that require jumpers for particular brand installations.

If there is the slightest reason to suspect the connections of a replacement part, you should first try the part out with temporary connections, such as clip leads, before mounting and soldering everything in place.

TO SUM UP: THE SUBSTITUTION METHOD IS USUALLY THE LAST STEP IN THE OVERALL TROUBLESHOOTING PROCEDURE. TO PERFORM SUCCESSFUL SUBSTITUTIONS, BE SURE TO USE THE CORRECT REPLACEMENT PART AND THEN BE DOUBLY SURE THAT IT IS ALSO CONNECTED CORRECTLY INTO THE CIRCUIT.

1.8   WHAT TO DO AS A LAST RESORT

Occasionally it appears as if a particular defect will defy every trouble-shooting technique and cannot be located at all. We usually come to this conclusion after hours of frustrating work. By this time we have to decide whether the defect is intermittent or whether the defect is really there all the time. If the defect is intermittent, i.e., it seems to be present at one time and seems to be absent another time, we know that a lengthy, special procedure will be required. Chapter 6 is devoted entirely to finding intermittent defects.

If the defect is not intermittent, but is so elusive that we cannot find it by any of the above four methods, the very worst thing to do is to panic or get excited. We have found that the best thing to do at this point is to stop the troubleshooting effort for the time being, to have a cup of coffee, smoke a pipe, go out for dinner, or somehow postpone the entire problem for a while, preferably until the next day. You will find that in the intervening period you will think about the defect off and on, and you will eventually realize just what you have overlooked, where the problem really is, and what you did wrong in general. When you then return to the workbench to troubleshoot the "impossible" defect, you are very likely to find the trouble relatively quickly.

If the defect is still elusive, the following procedure is recommended:

Write down, in key words, every troubleshooting step performed so far. Draw a simple schematic diagram of the circuit portion for which voltage and resistance checks have been made. List all of the voltages and resistances you have measured on this schematic. Compare the schematic you have drawn, with its measurements, with the manufacturer's data.

If the trouble is still a mystery, try these steps:

With the power turned off, use an ohmmeter or continuity tester to trace out every connection, pin to pin, wire by wire, from one junction point to the next, and compare it to the manufacturer's diagram. Pay particular attention to continuity between all points that the circuit diagram shows to be connected. Also, check the actual pin numbers on all sockets and connectors.

Another approach, which works well on TV and hi-fi equipment, is this:

Write down the symptoms observed. Now, look at the manufacturer's diagram and all of his service instructions and think of any kind of failure that could cause this type of defect. Consider the possibility that the symptom you have observed and the ones described in the manufacturer's literature do not appear exactly alike or are described in different words. (For example, picture is weaving, or diagonal stripes, really both due to loss of horizontal synchronization.)

For complex equipment like digital computers, the following is also recommended:

Looking at the manufacturer's circuit diagram, list all the other circuits that seem to operate correctly but that could affect the operation of the suspected section. Then, carefully test again how each of these circuits works, taking measurements to make sure that the manufacturer's specified signals really go to the suspected section.

Table 1-2 lists seven typical defects that can be so hard to find that all of the above troubleshooting methods seem to fail. The first item, poor solder connections, can often pass unnoticed, although visual inspection does sometimes show a poor solder connection. If there is the slightest suspicion that a poor solder connection could cause the defect, you can probably save yourself a lot of time and trouble if you simply apply a clean, hot, soldering iron to every single connection in the circuit section you suspect.

The second item, broken wires, would be obvious only when the two ends of the break are visibly separated. Very often, however, the break is invisible because it occurs either right at the point where the connection is made or else at the point where the insulation has been stripped off or, occasionally, within the insulation due to a very sharp bend. One method to find broken wires is to pull at every wire you suspect with tweezers or long-nosed pliers.

Partial short circuits occur whenever two conductors are close together and, in the absence of proper insulation, the partial short circuit may be caused by accumulated dirt, moisture, etc. Very often partial or temporary short circuits are due to the piercing of the insulation of a wire. This occurs particularly often in equipment where the wire-wrapping technique is used, when wires are guided around other wire-wrapped terminals at sharp angles and are under tension.

On PC boards open and short circuits are very likely and very hard-to-find trouble sources. Hairline cracks in the conductors can cause open circuits when the insulating material expands. Short circuits can

Poor solder connection
Broken wires
Partial short circuits
PC board open and short circuits
High voltage breakdown
Wrong component values
Wrong component connection

TABLE 1-2

be due either to dirt and moisture, or corrosion, or else they can be due to so-called "solder hairs," the bridging by fine solder strands. Open circuits on a PC board can only be found by continuity measurements and can be repaired by connecting an insulated wire across the two points served by the open conductor. Short circuits are difficult to find and it is often good practice to simply run a thin point, such as a scriber or similar tool, between conductors that are close together on the PC board.

High voltage breakdown occurs even at voltages as low as 100 volts, if a ground or negative voltage terminal is close enough. Again, dirt, moisture, and sometimes metal powders (typical for ferrite components) may form a bridge for high voltage breakdown. At the higher voltages used in transmitters and cathode ray tube devices (TV sets) high voltage breakdown is often due to corona, which is caused by sharp points at the high voltage area.

The last two items, wrong component values and wrong component connection, do not seem to be likely causes for trouble unless someone has already tried to troubleshoot the equipment and has installed the wrong components or the right component with the wrong connection, or both. In Table 1-1 we have asked, as the last question to the customer, whether anyone has tried to fix this unit before. Unfortunately, some people are too embarrassed to admit that someone else has already failed to find the defect and, when faced with a really tough problem, you might as well assume that someone else has been there before you. Look for components that are clean and new; look for connections that look different from the original manufacturer's process. Since even the best of us are only human and sometimes make stupid mistakes, you might also check whether any of the components that you have put into this equipment are the right values and whether they are connected correctly. We have replaced transistors two or three times on a simple amplifier until we realized that we had erroneously interchanged the emitter and collector wires. It took two days until we found this trouble and by that time we knew that there was nothing wrong with the original transistor, but that the defect was really due to a short circuit between a conductor on the PC board and the grounded case of a small electrolytic capacitor that was pressed against that conductor.

In some situations, when none of the troubleshooting methods yields any results, it is possible, without spending too much time, to take amplitude and waveform measurements with the oscilloscope of all of the signals in the defective circuit. If manufacturer's troubleshooting data are available, which contain these waveforms and amplitudes, a comparison with this data may provide a clue. If another equipment, operating correctly, is available, then it is possible to make comparison measurements between the defective one and the known good equipment. While this approach is relatively slow and cumbersome, and usually does not show up the actual defective component, it is very useful in tracing hard-to-define and unusual defects to a particular portion of the equipment.

TO SUM UP: REMEMBER, REGARDLESS HOW ABSURD AND IMPOSSIBLE IT SEEMS, EVERY DEFECT CAN BE LOCATED. BACK OFF, THINK, RELAX, AND THINK AGAIN. WITH THE STEPS DESCRIBED ABOVE YOU CAN LOCATE EVEN THE "IMPOSSIBLE" DEFECT.

2

Getting the Most from Your Test Equipment

2.1 WHAT THE TEST EQUIPMENT CAN DO FOR YOU

In chapter 1 we have seen that there are four basic troubleshooting techniques and we have learned that two of them, the symptom-function technique and the substitution technique, can be used with practically no test equipment.

In the symptom-function technique we have used the observable symptoms, as we can see and hear them, and combined them with the functions, as we understand the operation of the equipment, to isolate troubles to specific portions of the equipment. In the substitution technique, we simply substitute parts we believe to be good for those parts we suspect of being defective. It is often necessary to measure the electrical characteristics of suspected components, but the most frequent use of test equipment is in the signal-tracing and the voltage-resistance techniques. In both of these techniques we must have a means of measuring and observing the voltage, waveform, current, resistance, and other electrical parameters that control the operation of the equipment we are troubleshooting. Since electrical signals are invisible, test equipment is required to make them visible and to measure them. Electrical components may look okay, but test equipment is required to check their electrical characteristics. Test equipment is also required to provide test signals at selected amplitudes, frequencies, etc.

The basic question that confronts the electronics troubleshooter is to what extent he can rely on his test equipment. Experience shows that test equipment of good quality, high versatility, and accuracy is second in importance only to the troubleshooter's technical knowledge and understanding of the equipment. Good test equipment can save many man-hours, can avoid costly mistakes, and usually pays for itself in a relatively short time. This does not mean that an elaborate arrangement of many different kinds of equipment must be maintained, because your specific test equipment requirements depend on the type of troubleshooting work you specialize in. If you are in the television repair business, you will naturally work with test equipment specifically designed for this purpose. If you are doing digital computer troubleshooting, a different array of test instruments will be needed.

Specific equipment configurations will be discussed in the last three chapters of this book when specific types of equipment troubleshooting are discussed. Certain basic test instruments, however, are used everywhere in electronics. Regardless of which area you are working in, it is important to understand the basic operation of these widely used, common, and absolutely essential test instruments.

TO SUM UP: TEST EQUIPMENT IS NEEDED MOST FOR THE SIGNAL-TRACING AND THE VOLTAGE-RESISTANCE TECHNIQUES. IT IS ESSENTIAL FOR THESE TECHNIQUES AND KNOWING WHAT YOU NEED HELPS IN YOUR TROUBLESHOOTING WORK.

2.2 WHAT YOU SHOULD KNOW ABOUT YOUR TEST EQUIPMENT

Many electronics professionals, particularly if they are busy in their work, believe that it is sufficient to know how to read the meters and indicators of their test equipment. Frequently, when a new piece of test equipment is received, the instruction book is not even studied but only the immediately required functions are used to perform the test that is wanted at the moment. We often like to postpone a thorough study of the manufacturer's data for a later date, which never comes. As a result, test equipment that can perform many more functions is frequently underutilized or, in a surprisingly high number of instances, it is incorrectly used.

The professional electronics troubleshooter does not need to know all of the design details of the test equipment he is using, but he should be familiar with its operating principles and with all of its features, its limitations and the accuracies for different control settings. This information is readily available in the instruction book the manufacturer furnishes with his test equipment. In addition, specialized books provide a detailed knowledge of certain types of test equipment and often cover the application of the equipment in some detail. In the following paragraphs the basic operation of the most widely used, general-purpose, electronic test equipment is described and some typical circuit diagrams are presented. Toward the end of this chapter we will learn how to get the most from these test instruments, how to maintain and calibrate them, and how to use probes and leads effectively.

TO SUM UP: STUDY THE INSTRUCTION MANUAL FOR ALL FUNCTIONS. TRY YOUR NEW TEST INSTRUMENT ACCORDING TO THE INSTRUCTION MANUAL FOR ALL FEATURES AND FUNCTIONS. KNOWING ALL THE THINGS YOUR TEST EQUIPMENT CAN DO HELPS YOU IN TROUBLESHOOTING WORK.

2.3 BASIC OPERATION OF THE VOM, VTVM, AND OSCILLOSCOPE

As our readers undoubtedly know, VOM stands for volt-ohm-milliammeter. This simple and inexpensive device is probably the most widely used item in the whole range of electronic test equipment and no professional troubleshooter would consider doing anything at all without having a VOM at hand.

With all its popularity, however, the fundamental operation of the VOM is not always clearly understood. Figure 2-1 shows the three basic circuits used for the measurement of current, voltage, and resistance. To measure a current from a source V, having a source resistance R_S, the resistance of the meter movement itself, as well as the various switched-in shunts, R_1, R_2, and R_3, must be considered. It is clear from the basic circuit in Figure 2-1a, that the accuracy of the current reading will depend on several different factors. One is the question of whether the meter resistance remains the same over the entire range. The second factor is the accuracy of the shunting resistors when the different current ranges are used. The third factor is the effect of the measuring circuit itself upon the circuit on the test.

It is clear, for example, that the amount of current I flowing in the circuit will be determined not only by the voltage V and the source resistance R_S but by the shunting resistance, R_1, R_2, R_3, and the meter resistance itself. In order to obtain maximum accuracy, it is necessary that the resistance presented by the measuring circuit be much smaller than the resistance contained the circuit in which we are measuring the current.

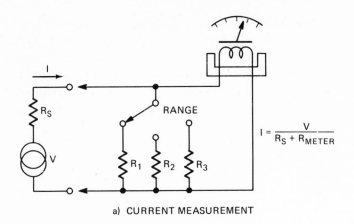

$$I = \frac{V}{R_S + R_{METER}}$$

a) CURRENT MEASUREMENT

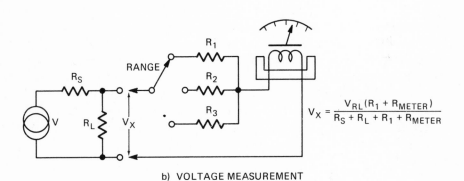

$$V_X = \frac{V_{RL}(R_1 + R_{METER})}{R_S + R_L + R_1 + R_{METER}}$$

b) VOLTAGE MEASUREMENT

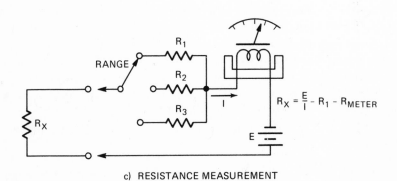

$$R_X = \frac{E}{I} - R_1 - R_{METER}$$

c) RESISTANCE MEASUREMENT

FIGURE 2-1
BASIC V.O.M. CIRCUITS

39

In the voltage measurement circuit in Figure 2-1b, the meter resistance is a relatively small factor. As illustrated, the voltage under test, V_X, should be determined only by the original source voltage V, the resistance R_S, and the load resistance R_L. Clearly, in this case, it is desirable to have a very high resistance value for R_1, R_2, and R_3. The amount of current flowing in the meter should be much less than 1/10 of the current flowing through R_L, if the voltage measurement has to have any meaning of accuracy. The formula shown together with Figure 2-1b explains the reason why VOMs are usually rated in terms of "ohms per volt." A typical value may be 5,000 ohms per volt and this means that when V_X in Figure 2-1b is indicated as 1 volt on the meter, the combined resistance of the meter itself would be 5,000 ohms.

The third circuit in Figure 2-1 shows what is required to measure an unknown resistance using a standard VOM. The main difference between the previous circuits and circuit (c), resistance measurement, is the introduction of a small battery E. Since the resistor under test is a passive device, this battery is necessary to drive a current I through the meter. The value of this current I is, of course, determined by the combined resistance R_X, R_1, and the meter resistance. R_1 and the meter resistance are known, the voltage across battery E is known and the only unknown element is, therefore, R_X.

In the three circuits shown in Figure 2-1, different ranges of current voltage and resistance measurement have been indicated by R_1, R_2, and R_3. Of course, in an actual circuit each of these resistances are of different values for the different measurements. In the case of the current measurements, for example, the smallest current can be measured when R_1 is infinite. The largest amount of current can be measured when the shunting resistor has a very low resistance. For measuring a low voltage, the series resistor R_1, R_2, and R_3 in the circuit (b) should be minimum while it should be very high for measuring higher voltages. For measuring resistance, a higher series resistor, R_1, R_2, and R_3, must be used for measuring a lower resistance R_X.

In an actual VOM, entirely different resistors, usually with some adjustment provided, are used for the different current voltage and resistance ranges. A typical example of such a circuit is shown in Figure 2-2. The circuit of the Simpson 260 VOM, probably the most

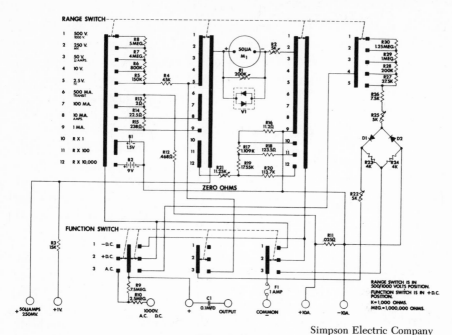

Simpson Electric Company

FIGURE 2-2
VOLT-OHM-MILLIAMMETER (SIMPSON 260)

widely used instrument of this type, shows a 12-position range switch and function switch. It also shows the detailed circuitry that makes up an actual instrument, including protective diodes across the 50 microammeter, a diode rectifier for measuring AC, and two separate batteries, B_1 and B_2, to cover the different ranges of resistance measurement.

One of the limitations of the VOM, as mentioned above, was the fact that the impedance of the meter itself affects the measurement of voltage. To overcome this limitation for circuits where small voltages and high impedances must be measured, the vacuum tube voltmeter was developed many years ago. The principle of this type of instrument is shown in Figure 2-3. Note that the only effective resistance shunting the load resistance R_L is R_g, the grid input impedance of the vacuum tube. In normal vacuum tubes this impedance can be quite high. Special circuitry can be used to increase this impedance further. The indication of the VTVM is provided by putting a milliammeter into the plate circuit and thereby measuring the plate current I_p. For all practical purposes, the voltage to be measured, V_X, controls the

41

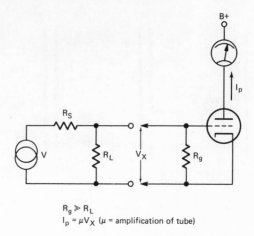

$R_g \gg R_L$

$I_p = \mu V_X$ (μ = amplification of tube)

FIGURE 2-3
PRINCIPLE OF VTVM

current in the plate circuit of the triode. Because of the amplification of the vacuum tube this makes it possible for relatively small values of V_X to have great effects on I_p and the meter reading.

Actual VTVMs usually used a dual triode, equivalent to the solid state version of the VTVM shown in its basic circuitry in Figure 2-4. Here the effective input impedance of the meter is the base impedance R_B. The indication of voltage is provided by a milliammeter, which measures the current due to the difference in collector voltage at Q_1 and

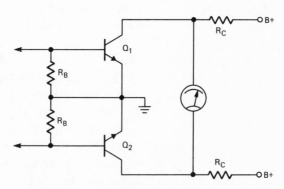

FIGURE 2-4
BASIC DIFFERENTIAL AMPLIFIER USED IN
SOLID-STATE VTVM

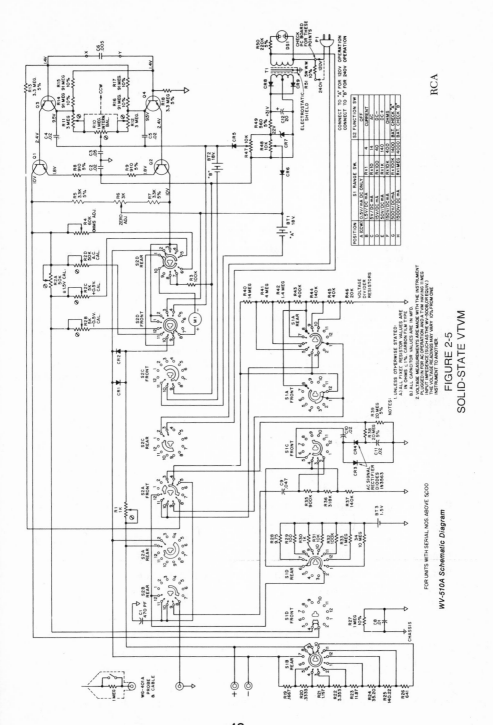

RCA

FIGURE 2-5

SOLID-STATE VTVM

WV-510A Schematic Diagram

FOR UNITS WITH SERIAL NOS. ABOVE 5,000

43

Q2. The basic difference between the regular VOM and the VTVM, using actual vacuum tubes or solid-state devices, is only in its voltage measuring circuit. The current and resistance measuring principles are the same, although VTVM-type instruments usually have a somewhat more sensitive meter movement.

The actual circuit of a typical solid-state VTVM is shown in Figure 2-5 and illustrates the circuit used in the RCA Master Voltohmist, shown in the photograph in Figure 2-6. Use of the probe and test leads in general is described later in this chapter.

The oscilloscope is the essential test instrument for observing the actual waveforms of signals. Figure 2-7 illustrates the basic operation of the oscilloscope as used in troubleshooting. Almost all oscilloscopes use electrostatic deflection cathode ray tubes and this implies that two sets of deflection plates control the appearance of any signals on the screen. A sawtooth sweep generator provides opposite polarity sawtooths, which are applied to the horizontal deflection plates. If no signal is applied to the vertical deflection plates, the result of the two sawtooth waveforms will be a horizontal line on the screen. As the spot generated by the electron beam is deflected horizontally across this screen, the vertical deflection plates cause this spot to move up and down in accordance with the voltage applied to them. By synchronizing the horizontal deflection with the vertical signal input, it is possible to show multiples of the signal under test. To observe a 60 Hz sinewave, for example, the sawtooth signal applied to the horizontal deflection plates could also be of 60 Hz frequency, synchronized so that the start of the sinewave coincides with the start of the horizontal deflection. Because the sawtooth wave applied to the horizontal deflection plates requires some, however short, time period to return the beam from the right back to the left side of the screen, it is not possible in this way to show a complete 60 Hz sinewave on the screen. For this reason it is common practice to set the sweep generator frequency to at least one-half of the signal frequency, and this would mean a 30 Hz sawtooth wave. This would put slightly less than two sinewaves on the screen.

The voltages applied to the deflection plates are usually on the order of several hundred volts. For this reason it is always necessary to provide amplification between the input leads of the oscilloscope and deflection plates, as shown in Figure 2-7. The amount of amplification, the linearity, and some calibration of gain, form important

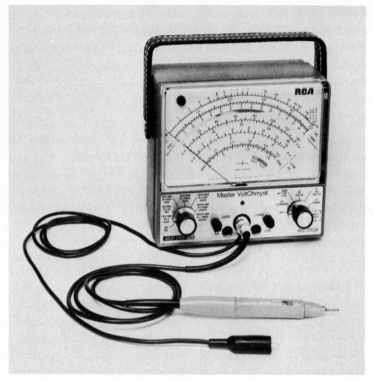

FIGURE 2-6

RCA

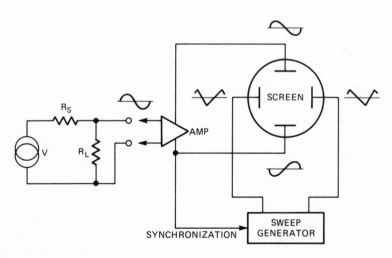

FIGURE 2-7
BASIC OSCILLOSCOPE OPERATION

characteristics of the oscilloscope. The frequency range that can be displayed by the oscilloscope depends on the bandwidth of the input amplifier as well as on the frequency range of the sweep generator. However, in order to display a 10 MHz sinewave, it may only be necessary to provide a sweep generator capable of producing saw-tooth waves at 1 MHz. Ten cycles of the sinewave would then be displayed on the screen.

Remember the basic operating principles of the oscilloscope: The way in which signals are applied to the deflection plates, the need for synchronizing the input of the vertical signal, the signal under test, with the sweep generator, the bandwidth of the amplifier and the amplitude or gain calibration of the amplifier. These are the essential characteristics that determine the utility of an oscilloscope for a particular troubleshooting task.

The block diagram of a typical service oscilloscope is shown in Figure 2-8. Beginning with the cathode ray tube, note that the vertical deflection plates are closest to the cathode and control grid. Because they are closer, the vertical plates will require a smaller signal for an equal amount of deflection than the horizontal deflection plates. High-voltage, focus, and intensity controls are provided as illustrated. Note that a separate stage, using transistor Q212, is used for blanking. This is necessary so that the retrace, the path of the electron beam from the right to the starting point at the left, is blanked out. A horizontal amplifier stage is used to provide the large sawtooth signals to the horizontal deflection plates. The section marked S403 is the synchronizing and horizontal frequency selection unit. This section receives a portion of the vertical input signal for internal synchronization on that signal. A separate input is available for synchronizing the horizontal sweep to some external source and a third input synchronizes the horizontal sweep to the 60 Hz power line. In a later section in this chapter we will see how useful these three separate synchronizing inputs can be. The actual sawtooth signal is generated in the sweep oscillator, according to the setting of the sweep range switch, S404, and the synchronizing signals. Because this oscilloscope is often used for TV work, a separate vertical and horizontal sync separator, Q206, is provided.

The vertical input, the signal under test, may be either AC or DC, selected by switch S401. An attenuator is followed by a preamp and amplifier, with two adjustments. One determines the vertical gain

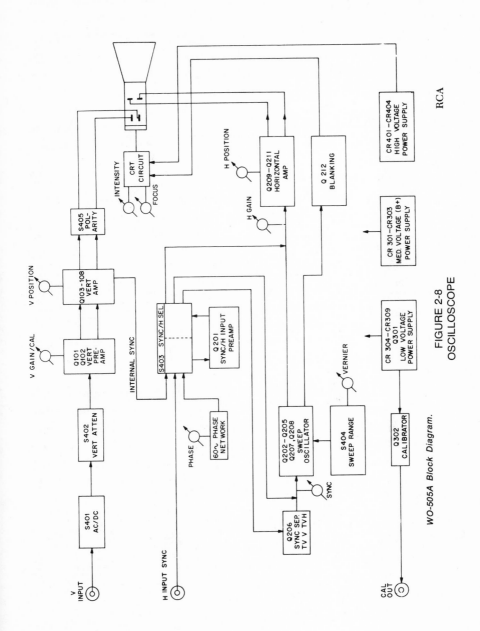

FIGURE 2-8
OSCILLOSCOPE

WO-505A Block Diagram.

RCA

while the other controls the position of the spot or line on the screen. Horizontal gain and horizontal position controls are also front panel adjustments.

Of particular value in an oscilloscope is the calibrator, Q302 in Figure 2-8. This circuit provides an AC voltage that can be used to calibrate the vertical amplitude of the oscilloscope and thereby make voltage amplitude measurements on complex waveforms.

TO SUM UP: THE VOLT-OHM-MILLIAMMETER ALLOWS YOU TO MAKE THE MOST ESSENTIAL MEASUREMENTS. REMEMBER HOW THEY WORK AND WHAT THE EFFECT OF THE METER CHARAC-TERISTICS CAN BE ON THE QUANTITY MEASURED. THE VTVM PERFORMS THE SAME BASIC FUNCTIONS BUT IS PARTICULARLY USEFUL FOR HIGH-IMPEDANCE CIRCUITS. THE PRINCIPLE OF OP-ERATION OF THE OSCILLOSCOPE DEPENDS ON VERTICAL AND HORIZONTAL DEFLECTION OF A CATHODE RAY TUBE, SYN-CHRONIZATION BETWEEN THE SIGNAL UNDER TEST APPLIED TO THE VERTICAL DEFLECTION PLATES, AND THE SAWTOOTH SWEEP VOLTAGE APPLIED TO THE HORIZONTAL DEFLECTION PLATES. THE OSCILLOSCOPE IS BASICALLY A METHOD TO "SEE" WHAT PERIODICALLY VARYING VOLTAGES ACTUALLY LOOK LIKE.

2.4 HOW TO GET THE MOST FROM YOUR METERS

The understanding of the basic operation of the type of meter you are using is the first step in helping you to get the most from this meter. When you understand how the meter works, you will use the correct settings of both the function switch and the range switch. You know that the meter's accuracy is generally best somewhere near the center of the scale and will therefore set the range so that the actual reading falls in the center.

What you measure determines what you read. If you are measuring DC voltage, be sure to read it on the DC scale. If you are measuring AC voltage, or AC current, be sure to read it on the correct scale since, although scales sometimes appear to be very similar, there is a good reason for the individual different scales on the meter.

Remember that almost all VOMs and VTVMs use a bridge rectifier at the AC setting. This means that the voltage indicated on the AC scale is usually in RMS for sinewaves, and, for square wave signals in average values. The difference between RMS values and peak-to-peak values (the maximum positive and the maximum negative excursion of the sinewave) is the factor of approximately 2.8. Some meters have peak reading scales and then it is necessary to know the frequency limitation for which these scales will hold correctly.

The polarity of the meter is very important for all DC measurements and is usually indicated on the meter terminals. The red test lead conventionally goes to the positive and the black test lead to the negative side. Polarity is not important for AC measurements but it is important when the ohmmeter is used to check electrolytic capacitors. Here, again, the positive meter lead should go to the side of the capacitor marked plus and the black lead should go to the opposite side.

On ohmmeter measurements we generally forget that the ohmmeter provides a current that has to pass through the resistance under test. Ordinarily this current is very small. In many VOMs, when the ohmmeter range is set to the lowest resistance value, as much as 100 ma can be passed through the resistance to be measured. When using quarter-watt and half-watt resistors, this is usually not a problem. When, however, the ohmmeter is used to test the resistance of diodes, particularly the base-to-emitter and base-to-collector diode junction of transistors, it is possible to burn out the transistor with the ohmmeter.

In addition to RMS and peak values, some meters also provide db and dbm scales for measuring audio signals. A zero dbm level refers to the audio signal of 1 milliwatt applied across an impedance of 600 ohms. The meter reading, in db, will vary when the circuit impedance varies. The dbm correction chart in Figure 2-9 shows the correction required when the circuit impedance changes from 600 ohms. For example, a 2.5 db correction must be applied in a negative sense if the impedance is 1,000 ohms and -12.5 dbm must be applied if the impedance is 10,000 ohms.

To check the calibration of your meter at the beginning of each working day, use a known voltage source, such as a new battery, and a known resistance, such as a 1% wire-wound resistor, to check the voltmeter and ohmmeter scales. In the case of the voltmeter, check the zero setting and, if necessary, correct it according to the instructions contained in the manufacturer's manual. For the ohmmeter it is

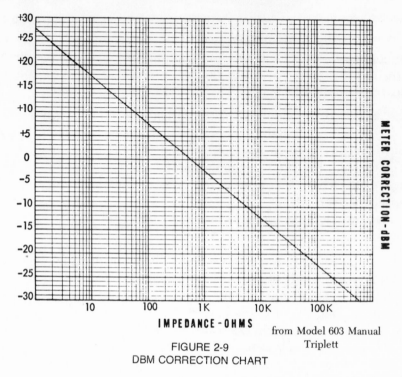

FIGURE 2-9
DBM CORRECTION CHART

from Model 603 Manual
Triplett

necessary to check both the "zero" and the "infinite" ends on the scale. To check for zero resistance, simply touch the two leads together and see that the meter shows zero for each of the ohmmeter ranges. Similarly, to check for the other end of the scale, simply separate the leads and see how far the meter needle indicates on each scale.

VOMs and VTVMs provide an adjustment to set the pointer to zero at the center of the scale. This feature is used for the alignment of the FM detectors and for a number of other, more specialized applications. Be sure to change .this setting back to the zero setting at the lower end of the scale whenever you change the function switch.

One of the principles of any measurement is to avoid changing the value that is being measured by the very process of measuring it. For voltage measurements this usually means that the impedance presented by the meter must be much greater than that in the circuit to be measured as was explained in section 2.3. Remember that the DC load and the AC load are not quite the same. Measuring AC voltage across a resonant circuit, for example, can produce a loading down of the circuit, even though the circuit would not be loaded down for DC.

Remember that the ohmmeter works from a battery and if it becomes difficult to set its pointer at zero or infinity, the battery, most likely, must be replaced. More detail on calibration of test equipment is given in paragraph 2.7 below.

TO SUM UP: USE THE CORRECT RANGE SETTING AND THE COR-RECT FUNCTION SETTING. KNOW WHAT YOU ARE MEASURING AND MAKE SURE YOU ARE READING ON THE CORRECT SCALE. CHECK POLARITY, CHECK ROUGH METER CALIBRATION.

2.5 HOW THE SCOPE CAN MEASURE ALMOST EVERYTHING

As we know from our basic understanding of oscilloscopes, the picture on the screen represents a plot of voltage versus time. It is therefore obviously possible with the oscilloscope to measure voltage and the time period between voltage changes. The basic characteristics that can be measured, and how these are measured are presented below:

AC voltages—Usually measured from zero to peak or from peak to peak. If an accurate vertical scale is provided on the screen, the RMS and the average values can be read off directly or else they can be computed from peak voltages.

DC voltages—Almost all oscilloscopes have a DC input setting. If the oscilloscope probe is connected to ground of the unit under test and the horizontal line on the scope is centered at zero, then, as the scope probe is moved to various DC voltages, the horizontal line will move up and down and its position can be read from the calibrated raster on the scope.

Frequency—Can be measured if the signal under test is synchronized with the scope's sweep generator. When two complete cycles of the signal appear on the scope, the frequency of the sweep generator will be one-half of the frequency of the signal under test.

Time periods—The sweep generator settings usually are in micro-seconds and milliseconds and permit direct reading of time on the horizontal scale of the oscilloscope screen.

Wave shapes—These can obviously be measured, just as in a paper and pencil plot, with voltage plotted vertically and time horizontally.

Phase angle—Phase angle can be expressed in time as well as in elec-trical degrees as a function of the frequency. The scope is particularly useful in comparing the phase angle of two signals of the same frequency. One signal is connected to the vertical and one to the horizontal scope input and this results in a series of patterns called "Lissajous figures." When both signals are in exact phase, a line describing a 45 degree angle between the horizontal and vertical will be shown on the scope. If the phase angle of the two signals is 90 degrees, a perfect circle or oval will be shown. More details on "Lissajous figures" can be obtained from any specialized oscilloscope instruction book.

Frequency response—This can be obtained if a sweep-frequency generator is used together with the scope. The sweep-frequency generator provides an RF or IF signal that varies in frequency with time. Usually a 60 Hz sweep rate, synchronized to the power line, is used to scan through the frequency band we want to observe. In TV servicing, for example, the IF frequency band from 41 to 47 MHz is scanned in this manner. The output of the amplifier under test is detected and fed to the vertical input of the scope while the 60 Hz time base is fed to the horizontal input of the scope. The resultant picture on the scope screen will then be a plot of frequency in the horizontal and amplitude response in the vertical direction. This technique is well known and will be described briefly again in the chapter on TV and FM service troubleshooting.

Impedance matching and standing wave testing—Can also be performed by means of a sweep-frequency generator and the oscilloscope. In this method the mismatch and reflections are shown similarly to the frequency response described above. Frequency is again plotted horizontally and the amplitude of the reflected or nonreflected signals are represented vertically.

The oscilloscope is such a versatile tool that many electronics troubleshooting professionals use it as their only, all-inclusive test instrument. Of course, you still need a VOM to measure resistance and current, but with a little ingenuity many in-circuit resistance tests can be performed with only the oscilloscope.

TO SUM UP: THE OSCILLOSCOPE LETS YOU "SEE" AND MEASURE ALL SORTS OF AC AND DC SIGNALS. STUDY THE INSTRUCTION MANUAL AND LEARN TO USE YOUR SCOPE FOR CHECKING PRACTICALLY EVERYTHING.

2.6 SIGNAL GENERATORS

In most instances where the signal-tracing technique is used, some sort of signal generator is required. Specific signal-generator applications will be discussed in chapters 8, 9, and 10 in conjunction with specific types of equipment, but all signal generators have a local oscillator that covers a certain frequency band. This signal can be modulated with audio, RF, video, pulse, etc., either by the amplitude (AM) or frequency modulation (FM). For TV work and occasionally for hi-fi troubleshooting, we use a sweep-frequency generator in which the frequency is periodically swept over a certain frequency band by electronic means. The rate of sweeping the frequency over the band is usually at 60 Hz and its application will be described in more detail in chapter 8. A third type of signal generator is the one that generates special signals for producing test patterns on color TV screens. More details on this type of generator will also appear in chapter 8.

Regardless of which type of signal generator is used, however, the following comments and suggestions will help you get the most out of the signal generator you need for your troubleshooting work.

Grounding—Between the signal generator, the equipment, and the meter or oscilloscope used in the signal-tracing method, is often the critical factor that causes wrong readings. In the case of RF, video, and pulse signals, it is important to move the ground point each time the "hot" portion of the signal generator lead is moved. In audio and low-frequency work, remember that the signal generator may have a balanced output, with a third terminal connected to signal-generator ground. Be sure to remember which kind of connection to the equipment under test is required.

Impedance matching—Be sure to know the output impedance of your signal generator. Typical values are 50, 75, or 600 ohms (balanced). If the input impedance of the circuit under test is 1,000 ohms and you shunt a 50-ohm signal generator impedance across it, this will substantially reduce the gain of that stage. Where necessary, impedance matching pads or transformers should be used. Where signal generators have output meters and calibrated output attenuators, remember that these readings refer to a perfect impedance match. In many circumstances, particularly at IF and RF, the impedance of the circuit under test will vary with frequency and this means that the impedance mismatch between the signal generator and the circuit under test will change as you tune the signal generator to different frequencies.

Output amplitude—Most signal generators are not required to maintain the same output amplitude at all frequencies and at all frequency bands. For this reason it is important to check the output amplitude each time the frequency is changed substantially. This can be done either by means of the internal meter or by using an external meter to check the generator output amplitude. In many cases the oscilloscope can be used to check the generator output amplitude as its frequency is adjusted. In general, the output amplitude will remain constant over the center portion of a given tuning band.

Modulation—Whenever the output signal is modulated, by a test tone, audio, video, or a pulse, the index or degree of modulation should be checked. One hundred percent modulation is generally undesirable in AM as well as FM. For FM modulation be sure to check the maximum deviation, which should be the same above and below the zero-modulation carrier.

Linearity—In some signal generators maximum output, just as maximum modulation, results in some distortion. This can usually be checked by the oscilloscope.

Instruction manuals—For some reason, the instruction manuals furnished with signal generators seem to be far more useful than those for other test instruments. It is a good practice to keep the instruction manual together with the signal generator. The instruction manual also invariably contains information on calibration and internal adjustments, which make the signal generator perform according to its specifications.

TO SUM UP: REMEMBER TO WATCH THE GROUNDING OF THE LEADS FROM THE SIGNAL GENERATOR TO THE EQUIPMENT UNDER TEST AND TO THE MEASURING EQUIPMENT. MATCHING IMPEDANCE AND CHECKING THE OUTPUT AMPLITUDE ARE IMPORTANT IN USING ANY KIND OF SIGNAL GENERATOR.

2.7 MAINTENANCE AND CALIBRATION

Most of the test equipment used for troubleshooting is designed for daily use on the bench, for taking along on service calls, and, generally, is able to withstand quite a bit of wear and tear. The accuracy required for most troubleshooting tasks is usually within 5% of the dial indications, and calibration of troubleshooting-type test equipment does not have to be as exacting as it would have to be for laboratory-type test equipment.

Most test equipment that we are likely to use in troubleshooting requires only very little maintenance. The only regularly replaced items are the batteries in the VOMs, but it is good practice to check and replace all worn leads, clips, probes, plugs, etc. It is also a good idea to tighten any loose screws, and to clean and lubricate the switch contacts of all switches in the test instruments. The AC line cords of those test instruments that we take on service calls are particularly vulnerable to damage and should be replaced frequently.

Most meters can be approximately calibrated at the center and both ends of the scale by following the instructions given in the manufacturer's literature and using the screwdriver adjustments inside the equipment. As voltage source we can use a fresh battery and a 1% wire-wound resistor can be used to calibrate the ohmmeter scales. Oscilloscopes can usually be calibrated against the 60 Hz power line at frequencies up to 1,200 Hz. Signal generators can be calibrated by having their outputs displayed on the oscilloscope. Both oscilloscopes and signal generators can be calibrated against a precision counter or against a special receiver capable of receiving WWV National Bureau of Standard calibrating broadcasts. This type of precision is not ordinarily required for test equipment used for troubleshooting.

TO SUM UP: REMEMBER TO CHECK AND REPLACE ALL WORN TEST LEADS, CLIPS, ETC. BASIC CALIBRATION FOR VOLTAGE AND RESISTANCE CAN BE DONE WITH BATTERIES AND PRECISION RESISTORS. CALIBRATION OF FREQUENCY CAN BE DONE AGAINST THE 60Hz POWER LINE OR AGAINST SOME OTHER ACCURATE PIECE OF TEST EQUIPMENT.

2.8 HOW TO USE PROBES AND LEADS EFFECTIVELY

Some kind of probe or test lead is furnished with every piece of test equipment. Sometimes the leads and probes are nothing more than insulated wires, but frequently a resistor, RC network, or a diode is located inside the test probe. We all are familiar with the DC probes used for VTVMs, which contain an isolating resistor and with the 10:1 attenuating probes used in oscilloscopes. RF probes frequently contain, in addition to the detector diode, frequency-compensating R-C networks. High-voltage probes contain very large resistance elements so that the original high voltage is divided down to something the VOM can read. The use of these probes is always described in the manufacturer's instruction manual and these probes should be used only for the purpose intended and not for other applications. Remember that most of the probes are designed not only to provide isolation, detection, or amplitude attenuation, but must also match into the impedance of the test equipment with which they are supplied. For this reason, it is really poor practice to switch probes around, such as between the oscilloscope and the VTVM.

One of the most versatile test instruments, and definitely the cheapest, is the clip lead. Experienced electronics troubleshooting professionals always keep a good supply of various clip leads on hand and they usually make sure that they have clip leads of at least two different colors, usually red and black. They also like to have the clips themselves protected by an insulating boot. Clip leads should be available in different lengths, with 6-inch, 10-inch, and 16-inch pairs of leads a good assortment. This permits you to make relatively short ground connections where necessary, and also allows you to reach to remote parts of a large chassis. Clip leads are handy to temporarily bridge suspected defective components or to short out certain points.

Figure 2-10 shows three uses of standard electrical components per-

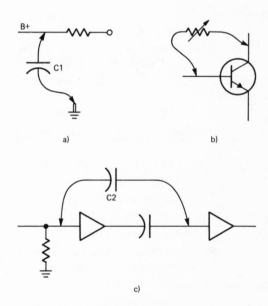

FIGURE 2-10
USE OF "CLIP-ON" COMPONENTS

manently soldered to clip leads. In Figure 2-10a, C1 may typically be a 100 MFD, 450 WV capacitor, which is used to connect any B+ points to ground to check for open filter capacitors. The potentiometer R in Figure 2-10b is very handy when determining the correct value of a resistor that has been burnt beyond recognition and for which we do not know the correct value. The capacitor C2 in Figure 2-10c is usually a smaller capacitor, .05 MF for audio work and .1 MF for video work, which can be used to short circuit signals to ground, to bypass suspected amplifier stages, and for many similar applications. Commercially available "substitution boxes" allow you to switch in components of different values once the circuit under test is connected by clip leads. This is very helpful when using the substitution method of troubleshooting.

TO SUM UP: ALWAYS KNOW WHAT A TEST PROBE IS SUPPOSED TO DO AND USE IT ONLY FOR THAT PURPOSE. CLIP LEADS ARE VERY HANDY FOR MANY APPLICATIONS—NEVER TRY A TROUBLESHOOTING JOB WITHOUT SOME CLIP LEADS IN YOUR KIT.

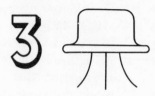

Troubleshooting Transistor Circuits

3.1 A BRIEF REVIEW OF TRANSISTOR FUNDAMENTALS

A full explanation of how transistors work, how they are used in different circuits, and how their performance is specified by characteristics, circuit equations, etc., is beyond the scope of this book. For those of our readers who need a brief review of transistor fundamentals, the following pages will help pave the way for the troubleshooting procedures presented later in this book.

Transistors come in many different sizes, shapes, and configurations, but they are basically all either of the PNP or the NPN type, depending on the arrangement of the semiconductor materials. The differences between the two types, especially concerning circuit polarity and the direction of current flow, is illustrated in Figure 3-1. Except for polarity of supply voltages and the current flow, the following discussion concerning PNP transistors is equally applicable to the NPN types.

Figure 3-2 shows the basic PNP transistor, together with the "Ebers-Moll" model that explains its operation. This model describes the transistor as two diodes, each of which is bridged by an amplifier with a gain of alpha. All of the equations concerning transistors can be derived from this model. Remembering the connection of the two diodes, it is also possible to test transistors, as explained later in this chapter.

Transistors can be operated in four different, basic configurations as shown in Figure 3-3. When both diodes are back biased, the transistor is cut off and no current flows, as illustrated in Figure 3-3a. When operating as a normal amplifier, the emitter junction is forward

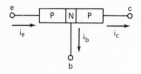

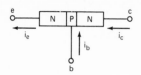

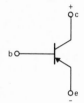

a) PNP TRANSISTOR

b) NPN TRANSISTOR

FIGURE 3-1
TWO POSSIBLE 3-ELEMENT TRANSISTORS

EMITTER BASE COLLECTOR

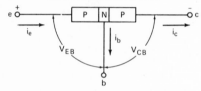

a) TRANSISTOR PARAMETERS

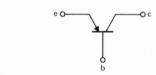

b) PNP TRANSISTOR

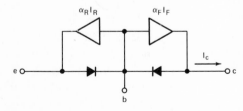

c) EBERS-MOLL MODEL

FIGURE 3-2
BASIC PNP TRANSISTOR

59

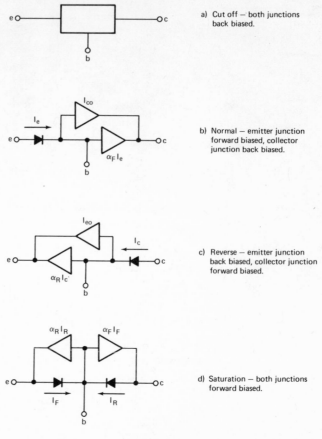

a) Cut off — both junctions back biased.

b) Normal — emitter junction forward biased, collector junction back biased.

c) Reverse — emitter junction back biased, collector junction forward biased.

d) Saturation — both junctions forward biased.

FIGURE 3-3
FOUR POSSIBLE TRANSISTOR CONFIGURATIONS

biased, the collector junction is back biased, and signals applied at the base appear amplified at the collector as illustrated in Figure 3-3b. This configuration is equivalent to a normal, grounded emitter, base input, collector output amplifier. Figure 3-3c shows the arrangement for an emitter follower or grounded collector connection. In this case, the emitter junction is back biased, the collector junction is forward biased, and input signals applied to the base appear at the emitter. The fourth possible configuration is shown in Figure 3-3d, when the transistor is said to be in saturation and both junctions are forward biased. In this circuit an increase in the base current will not cause any further increase in the collector or emitter current.

60

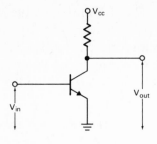

a) GROUNDED EMITTER

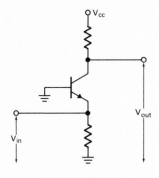

b) GROUNDED BASE

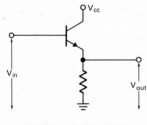

c) GROUNDED COLLECTOR
(EMITTER FOLLOWER)

FIGURE 3-4
THREE BASIC TRANSISTOR CIRCUITS

When used as an amplifier, a transistor can be connected in three basic circuits, as shown in Figure 3-4. The most common is the grounded emitter, in which the input signal is applied to the base and the output signal appears at the collector. For many applications where low impedance input is necessary, the grounded base config-

61

uration is used, particularly in RF circuits. The grounded collector or emitter follower circuit is used when a relatively high-input impedance, but low-output impedance, is required. In the grounded collector or emitter follower circuit, there is no voltage gain between input and output but, because of the impedance transformation, a power gain is realized.

A typical PNP amplifier circuit is shown in Figure 3-5 and illustrates the biasing requirements to assure that the transistor operates in true Class A. In this circuit, R_1 and R_2 are selected to cause some collector current to flow at all times. When the signal V_{in} is applied, the variation in the DC base current causes a corresponding variation in the collector current. In most circuits, R_E is either very small or else it is bypassed for the AC signal. For that reason, the output voltage is usually taken between collector and ground, but in order to be precise it should be measured between collector and emitter as shown.

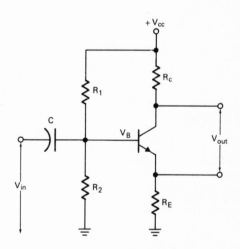

FIGURE 3-5
TYPICAL PNP AMPLIFIER CIRCUIT

Transistors are available in many different physical sizes, packages, and electrical characteristics. In audio and video amplifiers, it is common to classify transistors as small signal, medium power, and power amplifiers. In digital circuitry, transistors are classified according to the power that they can handle and according to the switching speed with which they can go from cut off to saturation and back to cut off.

Depending on the type of transistor and its application, different electrical parameters will be quoted in the manufacturer's data. One way of specifying transistors is by means of the so-called hybrid (h) parameters and another method relies on the alpha (short circuit current multiplication factor) and the beta characteristics. The beta characteristic equals $\frac{alpha}{1-alpha}$ and is therefore an indication of the current gain of a transistor. In addition to these characteristics, manufacturer's data usually states the maximum voltage to be applied between collector and emitter, the maximum collector current, emitter current, and the base current. It also states the maximum power dissipation for certain ambient temperatures for the transistor. For the troubleshooter, the characteristics of a transistor become very important when we try to replace a defective unit with one that is not an exact replacement type. Section 3.6 of this chapter deals with the limitations of the substitution techniques for transistor circuits.

The uniform scheme of numbering all semiconductors is limited to the first numeral. All semiconductors starting with 1N are diodes and all those starting with 2N or 3N are transistors of various types.

The type number of a transistor does not tell us whether it is a germanium or a silicon type, whether it is an NPN or a PNP, whether it is a switching transistor or a high-frequency amplifier, or, in most instances, whether it is an FET, MOS-FET, unijunction transistor, thyristor, SCR (silicon-controlled rectifier), or SCS (silicon-controlled switch). For this reason, one of the most important tools in trouble-shooting transistors is a handbook that lists all types and suitable replacements, together with at least the fundamental characteristics of each transistor. A typical manual may list as many as fifty different transistor package and pin arrangements, but the pin arrangements shown in Figure 3-6 are typical of the most widely used transistors. In each case the view is from the pin side.

3.2 HOW TO TEST TRANSISTORS

In the case of vacuum tubes, we can simply pull a tube out and substitute a new one or we can plug the suspected tube into a tube tester and see whether it is indeed good or not. Transistors are usually soldered into the circuit. In the few rare cases where transistor sockets are used, our experience has been that the sockets themselves are more likely to cause the defect, due to poor contact, than the transis-

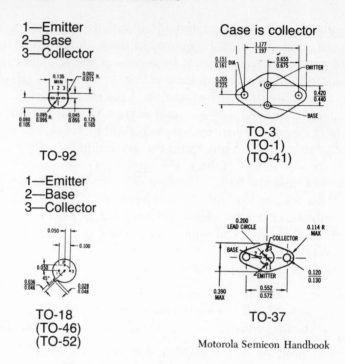

All transistors shown from pin side

FIGURE 3-6
TYPICAL TRANSISTOR PIN ARRANGEMENTS

tor itself. When we compare the reliability of individual transistors with other circuit components, such as capacitors, resistors, and inductors, the transistor frequently proves to be the most reliable component in the circuit. For this reason, we recommend testing the other circuit components before unsoldering the transistor and attempting to test it. Transistor testers are quite useful, particularly when troubleshooting equipment that plugs transistors into sockets, or when working on breadboard designs where transistors are easily unsoldered for test. Consult the manufacturer's manual for detailed instructions on how to use your transistor tester and be sure that the transistor type number is correct.

A simple ohmmeter can provide a great deal of information to the experienced troubleshooter. Referring back to the circuit in Figures 3-2 and 3-3, we remember that the transistor consists, basically, of two diodes. Since most transistors become defective by either a short circuit or an open circuit in one of those two diodes, diode testing of transistors is a relatively simple and easy way of determining whether a transistor is good or not. Before using the ohmmeter, however, consider the voltage and current that can be applied to the transistor on each of the different scales. In general, the highest resistance range might possibly apply excessive voltage and the lowest resistance range could possibly provide excessive current. For this reason, the middle scales, R x 10 or R x 100, are usually considered safe and most useful for testing transistors.

Figure 3-7 shows the ohmmeter connections and polarities for a PNP transistor. The base-collector diode tests, shown in Figure 3-7a, ver-

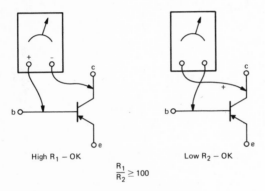

High R_1 — OK   Low R_2 — OK

$$\frac{R_1}{R_2} \geq 100$$

a) BASE-COLLECTOR DIODE TESTS

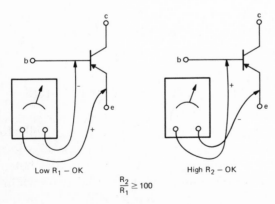

Low R_1 — OK   High R_2 — OK

$$\frac{R_2}{R_1} \geq 100$$

b) BASE-EMITTER DIODE TESTS

FIGURE 3-7
DIODE TESTING OF TRANSISTORS (PNP)

65

ify that the ratio of forward to backward resistance of the base-collector diode should be greater than 100. Similarly, the base-emitter diode tests shown in Figure 3-7b indicate the same arrangement, with different polarities, for the base-emitter diode forward and reverse ratio. As will be shown in section 3.6 below, more sophisticated tests can be arranged, but the four basic tests shown in Figure 3-7 are sure to locate well over 90% of all defective transistors.

As was pointed out at the beginning of this chapter, the two major transistor parameters are the alpha and the beta. For most trouble-shooting work we will not bother to set up test circuits to measure alpha and beta, but the transistor testers that are currently on the market use the principle shown in Figure 3-8. To measure alpha,

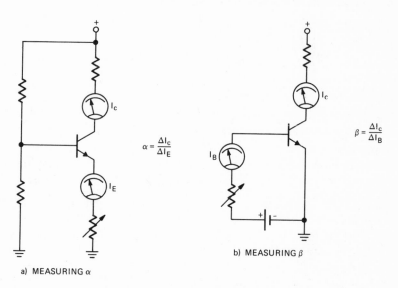

$$\alpha = \frac{\Delta I_c}{\Delta I_E}$$

$$\beta = \frac{\Delta I_c}{\Delta I_B}$$

b) MEASURING β

a) MEASURING α

FIGURE 3-8
BASIC TEST FOR a, β ON NPN TRANSISTORS

which is equivalent to the ratio of the change in collector current to the change in emitter current, for a given base bias level, we could use the circuit in Figure 3-8a. For measuring the beta, the current gain of a transistor, the base current is varied as shown in Figure 3-8b and the change in collector current is measured. The values of alpha and beta measured here should be within 10% or 20% of the values given in the manufacturer's data sheet, provided that all of the voltages used in the test circuit are exactly those specified.

3.3 HOW TRANSISTORS FAIL AND WHY

In theory, transistors can fail in many different ways, but in practice we find that defective transistors are either short-circuited, open-circuited, they are intermittent, or one of the essential electrical parameters has changed so that they can no longer be used in the original circuit. These failures can be caused either by an external mechanical influence, such as vibration or shock, or a fault in the manufacture of the transistor can limit its lifetime. Most frequently, transistors fail because of the failure of another component in the circuit. Because the transistor failure is so often connected with the failure of some other component, it is essential that, whenever a defective transistor is found, we carefully check all the other components that could affect the operation of the failed transistor.

Unlike the case of vacuum tubes, transistors, when properly manufactured and operated in the circuit, should last indefinitely. There is no cathode, no filament, and no other element to wear out. Manufacturing defects, however, are much more frequent in transistors than is commonly realized. The most widely found manufacturing defect, which causes transistor failure after many months of operation, is an uneven base element separating the collector and emitter elements. Any weakness in the base element will eventually cause a short between collector and emitter and, while this short circuit may be confined to a microscopically small area, it will eventually spread and cause an apparent change in parameters or else a catastrophic failure will occur, resulting in an emitter-collector short circuit. Other manufacturing defects, such as chemical contamination, may also slip through the manufacturer's test but may show up eventually. In these cases, a particular lot made on a certain day will show the same defect. Transistors used in military equipment undergo extensive temperature, shock, vibration, and altitude testing, so that the survivors of this test cycle are likely to be more reliable than the transistors that pass a minimal test and are then sold for commercial applications. Plastic-encapsulated transistors are often less reliable and more subject to manufacturer's defects.

Figure 3-9 shows the results of manufacturer's testing and subsequent failures of two transistor families, low-power silicon transistors, and medium-power (10 watts) silicon transistors. Note that in both types the short circuit mode of failure applies to 40%. The smaller transistors have a 53% incidence of parameter change, almost all of it due to

Failure Mode	0-1 Watt	0-10 Watt
Short circuit	40%	40%
Open circuit	7%	0
Intermittent	0	0
Parameter change	53%	0
Misc.	0	60%

FIGURE 3-9
SILICON TRANSISTOR FAILURES
AFTER MANUFACTURER'S TEST CYCLE

manufacturing problems. In the medium-power transistors, 60% of the failures are due to miscellaneous causes, such as defective glass seals, broken internal leads, loss of the seal, and defective heat-sink mounting.

Failures of transistors in military equipment, such as illustrated in Figure 3-10, show an entirely different pattern. Among the small-signal transistors, the majority failed due to parameter change, but now 31% show open circuit as against 7% of those found in the manufacturer's test cycle. In the medium-power range, open circuits and short circuits cause most of the defects.

These figures are of importance to the electronics troubleshooter because they tell us what type of trouble to look for, that is, how transistors are likely to fail and how these failures are easiest to find. Since short-circuit and open-circuit conditions account for almost half of all failures, it makes sense to test for short circuits and open circuits whenever a transistor is suspected. Testing for parameter changes is a little more involved but, as explained later, can also be done. In most cases, however, changes in transistor parameters can also be determined through the symptom-function technique, signal-tracing technique, and voltage measurements.

3.4 HOW TO USE THE SYMPTOM-FUNCTION TROUBLESHOOTING METHOD IN TRANSISTOR CIRCUITS

The symptom-function technique of troubleshooting depends on our understanding the functions of every circuit in the defective device and therefore works equally well whether the equipment uses integrated circuits, discrete transistors, or electron tubes. For this reason,

Failure Mode	0-1 Watt	1-10 Watt
Short circuit	8%	23%
Open circuit	31%	17%
Intermittent	3%	0
Parameter change	51%	5%
Misc.	8%	2%

FIGURE 3-10
FAILURES OF SILICON TRANSISTORS
IN MILITARY COMMUNICATIONS EQUIPMENT

detailed applications of the symptom-function technique are described in chapters 7 through 10 in connection with the specific equipment. Transistor circuits are unique, however, in the way in which transistors can become defective and this can be important when you use the symptom-function methods.

The typical failure modes of transistors were discussed in paragraph 3.3 above, but some practical examples of actual failures in the circuit will help us in using the symptom-function method more effectively. For example, open circuits in one or both of the diode junctions can be caused by a peculiar semiconductor phenomenon called "thermal runaway." This is due to the semiconductor characteristic of current gain increasing with temperature. As more current flows, the junction temperature increases, causing more current to flow, causing a further increase in temperature, and so on. Very frequently, thermal runaway is caused by the deterioration of another component, usually a resistor, which provides the necessary circuit resistance to prevent thermal runaway. The cure, then, is to not only replace the defective transistor but also to find and replace the other circuit element that caused the transistor to fail in the first place.

The effect of the failure of another circuit component on the apparent transistor failure is illustrated by the typical amplifier circuit in Figure 3-5. If resistor R_1 would be shorted, the transistor would always be in saturation and, depending on R_C and R_E, and on the power available, the transistor could reach a state of thermal runaway. Even if R_1 would only change in value, possibly due to aging, too much bias could be applied to the base of the transistor causing excessive current, saturation, nonlinear amplification, etc. If R_2 were shorted, the

69

transistor would be cut off and no signal would appear at the output. If R_2 were open, on the other hand, excessive current might again be drawn through the resistor due to the excessive base current. As this example shows, whenever we find a defective transistor we must make sure that at least the DC voltages and currents in the circuit are correct.

3.5   HOW TO USE SIGNAL TRACING IN TRANSISTOR CIRCUITS

The signal-tracing technique helps us locate defects down to a particular circuit regardless of the type of components, transistors, integrated circuits, or vacuum tubes used in the circuit. In a few respects, however, transistor circuits require special precautions when the signal-tracing method is used.

When the signal-tracing method was explained in chapter 1, we learned that in one version of this method, a known test signal is injected at a point of the circuitry and then traced through with an oscilloscope or meter. Because transistors have relatively low input impedance at their base, emitter and collector terminals, the point at which the test signal is applied and the level at which it is applied may often become very important. In the basic transistor IF amplifier shown in Figure 3-11, the impedances at points Ⓐ and Ⓒ are quite often different from the impedances shown at points Ⓑ, Ⓓ, and Ⓔ. Ⓐ and Ⓒ are the base circuits of the transistor that are returned, through a voltage divider, to the AGC bus. Points Ⓑ and Ⓓ are the collector circuits that are returned, through a resistor, to ground. The impedance at points Ⓐ and Ⓒ is determined, not by the voltage divider of the 2.7 K and the 12 K resistor going to the AGC bus, but by the transistor characteristics. This is apparent by the fact that the 680-ohm collector resistor, Ⓑ, is effectively shunted across the input impedance, Ⓒ, or the base of Q2. When connecting test signals from a generator to these points, we must know the output impedance of the generator and match it properly to the circuit. Connecting a 50-ohm generator, for example, to points Ⓑ and Ⓓ of transistor Q1 and Q2 respectively, would load down the resonant circuits made up of L1 and L2 respectively so much as to affect the frequency response of the IF amplifier considerably. If the 50-ohm generator, however, were applied at point Ⓐ or point Ⓒ directly at the base of the respective transistors, the loading

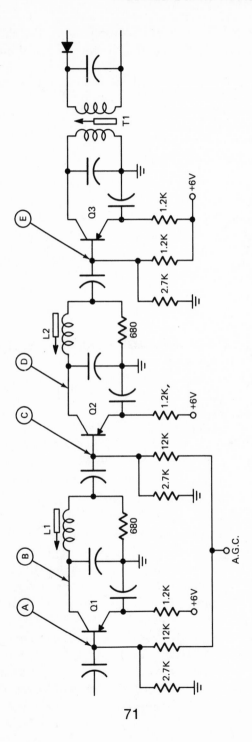

FIGURE 3-11
BASIC TRANSISTOR AMPLIFIER

71

effect on the tuned circuits would be minimized. For alignment of the output transformer T1, for example, the test signals should be applied to point Ⓔ , the base of Q3.

Entirely different problems will be encountered when using the signal-tracing method in troubleshooting digital transistor circuits or industrial controls. The basic principle of matching the impedance of the test-signal generator to the circuit impedance, however, is valid for all types of transistor circuits.

In many transistor circuits it is also important to limit the test-signal levels to avoid damaging the transistors. This is particularly important in the "front end" circuits of highly sensitive amplifiers where an excessively large signal can exceed the base-emitter or the base-collector breakdown voltage and cause a short circuit in the input transistor stage. If in doubt, check the transistor parameters in the transistor manual for the maximum base-emitter and base-collector voltage.

3.6 HOW TO USE THE VOLTAGE-RESISTANCE TECHNIQUE IN TRANSISTOR CIRCUITS

Transistors are basically current devices and this method of trouble-shooting could be called the "current-resistance" technique, but current measurements, especially in the circuit, are very cumbersome. Since we can use Ohm's law to convert current into voltage, once the resistance is known, the voltage-resistance technique, as described in chapter 1, can be applied to transistor circuits as well.

To illustrate how the voltage-resistance technique can be used in typical transistor circuits, refer to Figure 3-5, which shows a simple PNP amplifier. In this circuit, collector current will only flow in accordance to the amount of base-bias current. Instead of measuring base-bias current, we can measure V_{CC} and the voltage at the base, V_B. Assume that V_{CC} equals 12 volts, V_B 6 volts, and that R_1 is 10,000 ohms and R_2 20,000 ohms. Ohm's law tells us immediately that the current flowing through R_1 equals 12-6/10,000 = 0.6 ma. This means that 0.6 ma flows through R_1, but this is not the current that actually flows into the base of the transistor since R_2 also enters the picture. Six volts, again, is applied across R_2 and this means that the current through it is 0.3 ma. In other words, 0.3 ma flows through R_2 and 0.3 ma flows into the base of the transistor. If we repeat the procedure to

72

determine the collector current flowing through R_C, we can look up the manufacturer's data for the particular transistor and determine the operating point for the base and collector currents measured.

The above example illustrates that the voltage-resistance technique is not very handy for transistor circuits unless the manufacturer provides a chart listing the correct voltages and resistances at each point in the circuit. With such a chart we can compare the measured values against the specified values. For most simple transistor troubleshooting problems the voltage-resistance technique is used mostly to check suspected resistors, capacitors, and inductors after the defect has been traced to a particular circuit portion.

With troubleshooting equipment that has not been mass produced, but is either home-built or a kit, the voltage-resistance technique is frequently used to make sure that transistors operate at their specified current values. For these instances, Ohm's law is repeated here as a handy reference and reminder.

$$V = IR; I = V/R ; R = V/I$$

3.7 LIMITATIONS OF THE SUBSTITUTION TECHNIQUE IN TRANSISTOR CIRCUITS

In the vast majority of transistorized electronic equipment, the transistors and other components are soldered onto a printed circuit board that is sometimes encapsulated with a protective coating, making the removal of a transistor a major surgical operation. One of the great advantages of transistors is their small size, but this small size is one of the drawbacks in the application of the substitution technique. Special tools, such as microminiature soldering irons, tweezers, etc., are available to perform this type of repair work, but for troubleshooting purposes it is usually too difficult and time-consuming to replace transistors just to find out whether they are good or bad.

Experienced troubleshooters use some shortcuts that are reasonably effective, except in very complex, high-density microminiature assemblies. One of these shortcuts is to unsolder only two of the transistor's leads from the suspected circuit. The terminals of the replacement transistor can be tentatively tacked or clipped to the lands on the PC board. In attempting to substitute power transistors, we will find that the process of removing the suspected unit from a

2N1642–2N1731

TYPE	MATERIAL	POLARITY	REPLACE-MENT	REF.	USE	MAXIMUM RATINGS					
						P_D @ 25°C	Ref Point	T_J. °C	V_{CB} (volts)	V_{CE-} (volts)	Subscript
2N1642	S	P			SC	250M	A	160	30	6.0	U
2N1643	S	P			A	250M	A	160	25	25	U
2N1644	S	N	2N2218	2N2218	S	2.0W	C	175	60	40	R
2N1645	G	P			AHP	1.0W	A	100	35	20	O
2N1646	G	N			SH	150M	A	100	15	12	S
2N1647	S	N	2N3445	2N3445	AHP	267M	C	175	80	80	V
2N1648	S	N	2N3446	2N3445	AHP	267M	C	175	120	80	O
2N1649	S	N	2N3447	2N3445	AHP	267M	C	175	80	80	V
2N1650	S	N	2N3448	2N3445	AHP	267M	C	175	120	80	O
2N1651	G	P		2N1651	AP	100W	C	110	60	60	S
2N1652	G	P		2N1651	AP	100W	C	110	100	100	S
2N1653	G	P		2N1651	AP	100W	C	110	120	120	S

ELECTRICAL CHARACTERISTICS									
h_{FE} @ I_C (min)	(max)	Units	$V_{CE(SAT)}$ @ I_C (volts)	Units	h_{f-}	Subscript	f_-	Units	Subscript
15		100*							
10	25	100*							
40	120	150M	1.5	150M			50M	T	
20		0.2A	4.0	0.2A	25	E	450M	T	
20		10M							
15	45	0.5A	3.0	1.0A			3.0M	T	
15	45	0.5A	3.0	1.0A			2.0M	T	
30	90	0.5A	3.0	1.0A			3.0M	T	
30		0.5A	3.0	1.0A			2.0M	T	
35	140	10A	0.65	25A	20	E			
35	140	10A	0.65	25A	20	E			
35	140	10A	0.65	25A	20	E			

Motorola Semicon Handbook, 5th Edition

FIGURE 3-12
TYPICAL TRANSISTOR AND REPLACEMENT CHART

heat sink may be lengthy and involved. The substitute transistor often requires a heat sink even for a short test run and experienced servicemen sometimes simply use a clamp or tape a large piece of copper to the temporary substitute while disconnecting two leads from the suspected unit.

Aside from the mechanical problems in using the substitution technique, the vast variety of transistor types makes the probability of having the exact replacement type on hand very poor. Figure 3-12 shows an excerpt of a typical manufacturer's guide to transistor replacement. The left columns list the transistor type numbers, whether it is an NPN or PNP, germanium or silicon. For some of these transistors there is a specific replacement type. For others, the

Material: S = Silicon   G = Germanium
Polarity: P = PNP   N = NPN
Use: A   =   Amplifier
            AHP =   Amplifier, high frequency, power
            AP   =   Amplifier, power
            S   =   Switch
            SH   =   Switch, high speed
            SC   =   Switch, chopper
PD = Power dissipation, M = milliwatt, W = Watt
Ref. Point: A = Ambient, C = Case, J = Junction, S = Stud,
TJ = Max. operating junction temperature
Subscript (Refers to VCE max):
            U   =   VCE, Termination undefined
            R   =   VCER, Specified resistance
            O   =   VCEO, Base open
            S   =   VCES, Base shorted
            V   =   VCEV, Used only with voltage bias
Hfe =   Common emitter, DC short circuit forward current transfer ratio
Hf =   Small signal forward current transfer ratio
Subscript: Defines parameter, e.g. E = Hfe
f =   Cut-off frequency
Subscript: Defines parameter, e.g. T = fт = Current gain—bandwidth

FIGURE 3-13
EXPLANATION OF FIGURE 3-12

electrical characteristics are so close that they can be used, regardless of the pin connections. The column headed "REF" indicates the transistor type numbers for which the electrical characteristics are referenced in detail. Note that transistor type 2N1648 can be replaced by 2N3446. We can look up the electrical characteristics of 2N3446 under 2N3445 indicating that these latter two have the same characteristics.

A summary of other useful information is contained in the rest of the columns. Figure 3-13 contains an explanation of each of the characteristics shown in these columns. If we just go over some of the characteristics that are concerned here, and the actual numbers and the differences from one transistor to the next, we can readily see the problem involved in using anything but the exact, original, transistor type number as a substitution.

Most experienced troubleshooters use the substitution technique in transistor circuits only as a last resort, only when the defect has been isolated to a particular transistor and when the voltage-resistance measurements do not clearly show just what the defect is. Substitution of transistors is really effective in cases where the transistor itself contains an intermittent defect or where its temperature characteris-

75

tic is definitely outside the specified limits. In such cases, we find that substituting a "known good" transistor of exactly the same type cures the problem.

3.8 TYPICAL TRANSISTOR TROUBLES

In paragraph 3.3 above, we have described how transistors fail and why. This description included primarily the normal failure modes and the likelihood of certain types of failures in transistors operating in the equipment. The following paragraphs describe typical transistor troubles encountered by the author and by his friends in recent troubleshooting efforts.

(a) The power output transistor in a hi-fi system is found to be "burned out"—both diodes appear open. A new power transistor is installed, the equipment works fine, but within a week the power transistor is again "burned out." When we measure the voltage at the base of this power transistor, while the unit is heated with a hair drier, it turns out that the driver transistor base bias changes, causing the driver transistor to saturate and this, in turn, causes thermal runaway in a power transistor. The base bias of the driver transistor was determined through a combination of potentiometer and diode. By replacing the diode, which tested okay at room temperature, the trouble was cured.

(b) An intermittent IF amplifier in a color TV set was traced to a piece of solder, apparently left over from a previous troubleshooting job, which caused an intermittent short circuit between two conductors on the IF PC board assembly.

(c) In an industrial digital timer device, a shorted transistor was replaced. The new transistor, which had tested perfectly good before being installed, was shorted again. Three transistors were replaced and each was found to be defective even before AC power was applied. It was found that the soldering iron used in replacing each transistor had a 40-volt leakage between the AC power element and the soldering iron tip. The PC board was grounded to the AC power line because the power cord had remained in the socket, and an effective 40 volts was applied across each replacement transistor as we soldered it in place. Naturally, this blew the transistor. Ever since then we only use low-voltage soldering irons, which include an isolation transformer from the power line.

(d) One of the control transistors in a regulated power supply was found to be shorted and was replaced. The power supply operated apparently correctly but the range of regulation was severely lim-

76

ited. We replaced the new transistor twice. After a lot of useless attempts at analyzing the circuit, it was finally determined that when the first transistor had been replaced, we had inadvertently connected the collector lead to where the emitter should have gone and vice versa. Because of the particular circuit arrangement, this did not directly affect the power supply output, but it did not permit this transistor to function properly in the control circuit.

From these and many other instances of troubleshooting transistor circuits, we have learned a few precautions that apply to all work with transistors. Whenever new transistors have to be installed, remember that transistors, particularly the germanium types, are very sensitive to heat. Always pretin the leads while applying a heat sink, such as an alligator clip, between the tinned portion and the case of the transistor. Be sure to use a soldering iron that is isolated from the power line and operates at a temperature not exceeding 600° F. If at all possible, when the transistor leads are soldered in place, hold the tips of a needle-nosed pliers between the soldering iron and the body of the transistor to act as heat sink. Be sure to avoid sharp bends on the transistor leads too close to the case to avoid damage to the glass seal. It is a good idea to do this even on epoxy or other plastic transistors to avoid cracking the plastic where the lead enters it. In testing transistors remember, as pointed out in paragraph 3.2 above, that the voltage and the current applied to the transistor must be limited to within the transistor's ratings. Review paragraph 3.2 for the proper test procedure when using an ohmmeter.

Troubleshooting transistor circuits requires a knowledge of the circuit, but it also requires very careful handling of the transistors themselves. Unless you make sure that you do not damage the transistors in the course of the troubleshooting or replacement, your troubleshooting effort will be many times as complex and expensive as it needs to be.

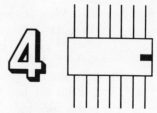

Troubleshooting
Integrated Circuits

4.1 A BRIEF REVIEW OF IC FUNDAMENTALS

If you understand how diodes and transistors work, then you should
have no difficulty understanding integrated circuits (ICs), because
they are merely the functional transistor and diode circuits produced
by somewhat different processes and assembled in a very small space.
All the circuit elements, the semiconductors as well as the resistors
and capacitors, are all contained on a single solid-state chip and some
ICs contain as many as several hundred transistors and resistors in a
single package, not more than ¼ inch x ¼ inch square.

Two different methods of construction are used to make ICs. The
so-called monolithic process means that all of the components,
semiconductors, resistors, capacitors, etc., are manufactured at the
same time, on the same substrate, and operate as completely func-
tioning circuits. In a second method, called hybrid construction, different portions of the circuit are manufactured separately and are
then assembled on the substrate, which also provides the interconnection paths. The whole hybrid circuit is then packaged and, as far as
the user is concerned, it operates basically the same as a monolithic
circuit. Hybrid construction is used usually when capacitors larger
than .001 MFD or precision resistors and inductors are required to
make up a circuit. These elements are produced as separate chips,
which are then assembled and interconnected on the main substrate.
The substrate contains the interconnecting pattern, similar to a
printed circuit, but many times smaller. As a rule, hybrid circuits are
custom-made for particular functions, while monolithic circuits are
available as standard off-the-shelf items, somewhat like transistors.
Monolithic circuits are used almost exclusively in digital logic circuits

for computers and other digital devices, while hybrid circuits are generally used for analog circuits and special functional assemblies.

Digital logic circuits form the majority of all ICs and they are the ones that are now available in so-called medium-scale integration (MSI) and large-scale integration (LSI). These two techniques mean that, instead of a single circuit, a large number of circuits is contained on a single chip and is packaged as a functional unit. A small-scale integrated circuit, for example, will contain four 2-input gates. An MSI may contain a complete binary to decimal decoder. An LSI, one of the most recent types, may contain all of the logic circuits necessary to form a complete pocket calculator.

Digital logic circuits are available in four different families, according to the circuit elements that they use to perform the logic function. As shown in Figure 4-1, these four basic circuits are the resistor-

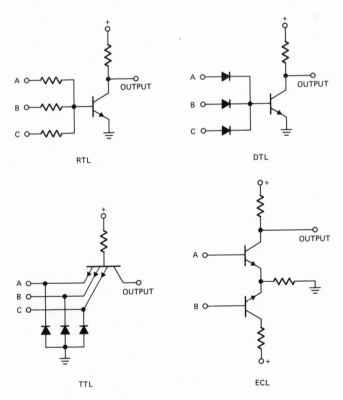

FIGURE 4-1
BASIC DIGITAL LOGIC FAMILIES

79

transistor logic (RTL), the diode-transistor logic (DTL), the transistor-transistor logic (TTL), and the emitter-coupled logic (ECL). The basic circuits shown in Figure 4-1 illustrate the principles of each family type, but in the actual IC, output stages are added to provide sufficient drive capability to drive a number of other logic circuits.

Figure 4-2 shows the circuitry contained in a typical flip-flop, together with the pin connections for the dual in-line package containing two such circuits. An MSI configuration using this type of flip-flop as many as eight times, together with some additional logic gates and inverters, is shown in Figure 4-3, which shows an 8-bit shift register contained in a single IC package.

The four logic families described above all use the same basic junction transistors. In recent years new families of transistors, based on different principles, having considerably different characteristics, have been available. The field effect transistor (FET), the metal oxide semiconductor (MOS), and the combination MOS-FET, are semiconductors that operate very much like transistors. FETs use a considerably higher voltage, less current, and much more isolation between the gate or base element and the main conducting elements, the collector and emitter, which are called the source and the drain. The latest of the low-current devices is the complementary MOS (CMOS), and the so-called charge-coupled (CC) LSIs. Space does not permit us here to go into detailed discussions of these devices, but when you encounter them you simply have to look up the characteristics knowing that, fundamentally, they operate the same way as transistors but they use much less current and are particularly useful for certain high-frequency applications.

The majority of analog ICs are either differential amplifiers or operational amplifiers. There is another branch of analog ICs, which could also be considered digital, because they act as analog gates and use either the FET or the MOS-FET construction. These gates are controlled by digital signals but act like switches in that they present either a very high or a very low series resistance for analog signals. FET gates are typically used for switching audio signals, controlling the input to amplifiers, etc.

A typical IC operational amplifier (op-amp) circuit and its equivalent circuit is shown in Figure 4-4. Note that thirteen transistors, fifteen resistors, and two diodes make up this amplifier, all manufactured together and contained on a chip approximately ⅛" x ⅛" square. The

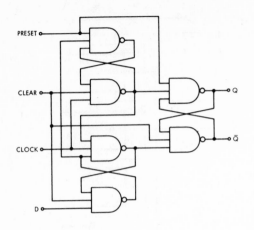

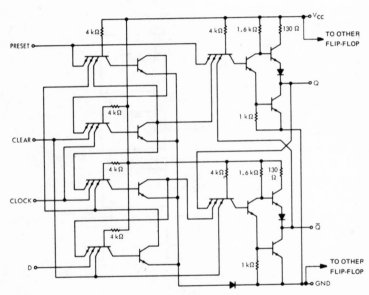

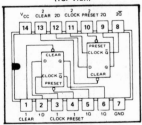

J OR N DUAL-IN-LINE PACKAGE
(TOP VIEW)

Texas Instruments

FIGURE 4-2
TYPE SN7474 DUAL D-TYPE FLIP-FLOP

81

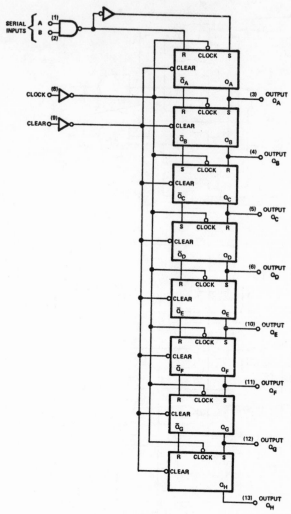

Texas Instruments (SN54/74164)

FIGURE 4-3
8-BIT SHIFT REGISTER IC

From *Buchsbaum's Complete Handbook of Practical Electronic Reference Data*, Walter H. Buchsbaum, ©1973, Prentice-Hall, Inc., p. 214.

chip, of course, is mounted in a larger package such as a TO-5 can. In order to use this operational amplifier in a circuit, it is usually necessary to use some external components as well. Figure 4-5 shows a typical narrow-band IC amplifier with its external components and using an op-amp chip similar to that shown in Figure 4-4. Note that in

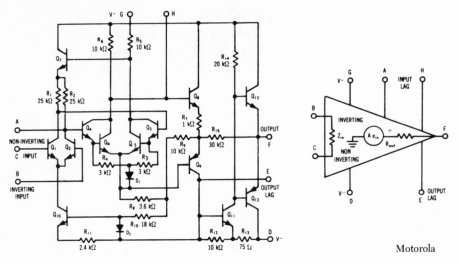

Motorola

FIGURE 4-4
CIRCUIT AND EQUIVALENT OF MC1709C OP-AMP

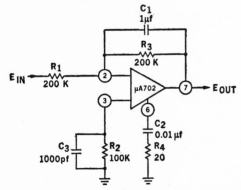

Fairchild Semiconductors

FIGURE 4-5
NARROW-BAND AMPLIFIER WITH EXTERNAL COMPONENTS

From *Buchsbaum's Complete Handbook of Practical Electronic Reference Data*, Walter H. Buchsbaum, ©1973, Prentice-Hall, Inc., p. 208.

Figure 4-5 each of the external control circuits contains a relatively large capacitor, C_1, C_2, and C_3, which cannot be manufactured as part of the monolithic process. By changing the values of these capacitors and their associated resistors, the bandwidth, time constant, etc., of the total circuit is controlled.

83

From our point of view as troubleshooters, the detailed circuitry inside the IC package is of little interest since we cannot possibly repair it or even get to it. For troubleshooting, therefore, the outside package, the leads, and the way we can get to these leads are the most important aspects.

Three basic IC packages are in widespread use. Relatively few commercial applications use the metal cans, similar to the transistor case, ranging from the TO-5 to TO-99 to larger variations of the same can. ICs mounted in cans are usually limited in the number of pins that are available. Many military equipments, particularly where space is at a premium, use the so-called "flat pack" approach. As the name implies, these are rectangular packages with less than 1/16-inch thickness, with flat leads coming out at the sides. Flat packs are invariably soldered on to a PC board or other substrate by a machine-soldering process. The most widely used package for ICs is the dual in-line package (DIP), which comes in standard 14- and 16-pin configurations. To accommodate the much larger number of pins for MSI and LSI, these ICs are packaged in larger DIP packages with 24 to 40 pins available. While the majority of DIPs are soldered into PC boards, in some equipments and, of course, in most breadboard equipments, DIP sockets and plug-ins are used. Figure 4-6 shows the approximate appearance and size of the most popular IC packages.

Once you have tried to use test clips and probes on any kind of IC package or configuration, you will understand why we claim that access and connection to these pins is the main problem in troubleshooting ICs. A number of manufacturers have come up with solutions to this problem. Typical of these is the Pomona Electric Company clip, which connects to a standard 14- or 16-pin DIP. The obvious solution to the problem of connecting to pins on the IC would be to connect to other suitable points on the PC board. As we will see later in this chapter, when we learn how ICs fail, presence of a particular voltage or signal at the test point on the PC board does not necessarily mean that this same signal really reaches the IC. Pressing a sharp point against the connector pin of a typical DIP or flat pack is not a very good idea either because any excess pressure from the test probe may cause a total or at least intermittent failure at the point where the pin enters the IC case. Most equipment manufacturers are aware of this problem and provide test points for signal tracing and for voltage and resistance measurements on the PC boards, which give at

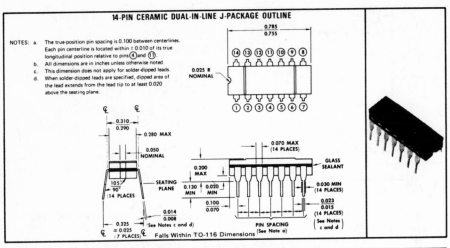

14-PIN CERAMIC DUAL-IN-LINE J-PACKAGE OUTLINE

NOTES: a. The true-position pin spacing is 0.100 between centerlines. Each pin centerline is located within ± 0.010 of its true longitudinal position relative to pins ④ and ⑪.
b. All dimensions are in inches unless otherwise noted.
c. This dimension does not apply for solder-dipped leads.
d. When solder-dipped leads are specified, dipped area of the lead extends from the lead tip to at least 0.020 above the seating plane.

Falls Within TO-116 Dimensions

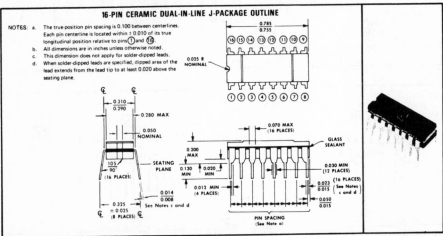

16-PIN CERAMIC DUAL-IN-LINE J-PACKAGE OUTLINE

NOTES: a. The true-position pin spacing is 0.100 between centerlines. Each pin centerline is located within ± 0.010 of its true longitudinal position relative to pins ① and ⑯.
b. All dimensions are in inches unless otherwise noted.
c. This dimension does not apply for solder-dipped leads.
d. When solder-dipped leads are specified, dipped area of the lead extends from the lead tip to at least 0.020 above the seating plane.

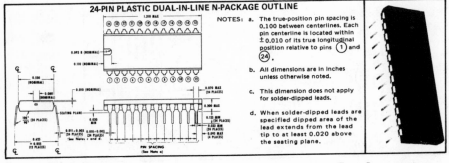

24-PIN PLASTIC DUAL-IN-LINE N-PACKAGE OUTLINE

NOTES: a. The true-position pin spacing is 0.100 between centerlines. Each pin centerline is located within ± 0.010 of its true longitudinal position relative to pins ① and ㉔.
b. All dimensions are in inches unless otherwise noted.
c. This dimension does not apply for solder-dipped leads.
d. When solder-dipped leads are specified dipped area of the lead extends from the lead tip to at least 0.020 above the seating plane.

Texas Instruments

FIGURE 4-6a
TYPICAL IC PACKAGES

85

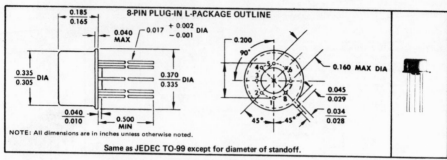

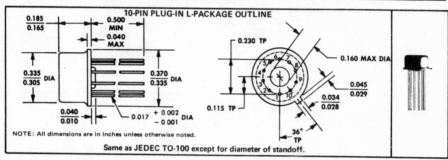

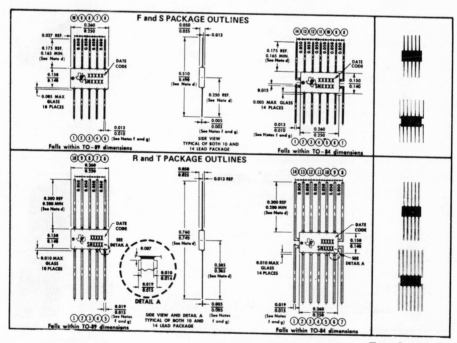

FIGURE 4-6b
TYPICAL IC PACKAGES

Texas Instruments

86

least a fair idea of the status of these ICs without requiring that the test probe be put directly on the IC lead. In any event, a description of typical IC troubles, at the end of this chapter, will give you a better idea on how to cope with this problem.

4.2 HOW TO TEST ICs

In chapter 3 we have discussed the problem of testing transistors. We found that transistors could be tested for short and open circuits relatively simply by checking the forward and reverse resistance of each of the two diodes that make up a transistor. When dealing with ICs this is not possible because, for one thing, we cannot get at the collector, base, and emitter of each of the transistors inside the IC package. Checking a single transistor would not be sufficient since open and short circuits in the interconnection pattern that makes up the whole IC can be just as troublesome as defective transistors or diodes.

Just as there are transistor testers available, so are there some IC testers on the market, but their applications are limited. One reason is that, just as in the case of transistors, it is necessary to remove the suspected IC from the circuit board and this usually means unsoldering. While a transistor has only three connections, the most frequently used ICs have fourteen or sixteen pins. It is quite a job to unsolder all these pins and remove the IC cleanly from the PC board even if you have one of the special IC unsoldering tools. Then you have to clean off all of the fourteen or sixteen pins, and make sure that they are straight so that they can be plugged into the socket of the IC tester. We do not know of any troubleshooter who really uses IC testers to a large extent. Most of the electronics technicians dealing with equipment using ICs in considerable quantities depend on entirely different approaches in testing ICs.

Whenever external components are used with ICs, the first and simplest approach is to make sure that these components work correctly. This is particularly important in analog circuits where a change in value of a feedback resistor can change the entire characteristic of the amplifier. For a general, overall approach to troubleshooting ICs, a most practical way is to check for the power supply voltage, to make sure that it appears at the proper terminal at each IC and to get some rough idea of the current drawn by that IC. Before actually measuring

87

the current we can check, just by touching it, whether it draws too much current—whether it gets too hot. Any time that an IC is too hot to keep your most sensitive finger on it for more than a few seconds, either the IC or some associated part is defective. Actual temperature measurements can be made with thermometers that measure surface temperature or with "Tempi-Lac," a paint that changes color at certain temperatures.

Digital ICs are usually operated from one or two fixed voltages and, if these voltages are within ten percent of the meter reading, we can assume that the operation of the ICs themselves is not affected by the supply voltage. Any analog ICs, however, are quite sensitive to their supply voltage. This is particularly true of those analog ICs that use control voltages, bias, etc., in addition to the V_{CC} supply. Whenever you suspect a defective analog IC, it is therefore important to carefully measure all voltages coming to this unit. Be sure to use a meter that is accurate to at least plus or minus 5%. Check the manufacturer's handbook for the voltages recommended for each particular IC and also check the circuit diagram of the equipment manufacturer for any notes regarding voltages.

LSIs generally are much more expensive than SSI ICs and for that reason, before removing an LSI from the circuit on the suspicion that it is defective, we must be particularly careful to check all of the voltages and external components controlling its operation. Once we are sure that the defect must be within the IC, it must, of course, be replaced with a new one. Be sure, however, that the voltages and currents are accurately measured and that at least all of the input signals are present in their correct waveforms and amplitudes. In terms of the systems concept, we must be sure that the input is exactly as it should be before suspecting the black box of giving us the wrong output.

The ideal way of testing ICs is in the circuit, with access to all of the pins. Hewlett-Packard has announced their 5011-T troubleshooting kit, which can be used for DTL and TTL-DIP ICs. Although this kit consists of several components, requires considerable skill to use, and is limited to testing those two logic families, it is priced at $625, more than most electronics troubleshooting technicians are likely to pay. The kit contains a logic comparator in which a special multipin clip is attached to the suspected IC. Then a second IC of the same exact type is plugged into the comparator box and sixteen light-emitting diodes

(LEDs) indicate the difference on a pin-by-pin basis between the known good or new IC and the one in the circuit. Another part of the kit contains two probes, one generating the signal pulses and the other one indicating the high or low state of the circuit. With these two probes it is possible to signal-trace through individual logic stages. Additional refinements include a clip combination, which provides both stimulus and response testing for in-circuit ICs. This troubleshooting kit is quite handy and worthwhile for those who specialize in the area of TTL and DTL logic work.

For general troubleshooting, the best test method for ICs uses the systems concept, considering the IC as black box. If you know what the input is and if you know what the output should be, you can tell by measuring the input and output whether or not the IC is good. As illustrated in Figure 4-7, this principle can be applied to an FM-detector IC such as is commonly found in the audio portion of TV receivers. We can check, with a wide-band, high-gain oscilloscope, whether there is an input signal to the FM detector at terminals 1 and 2. We could also check for the presence of this signal by means of the

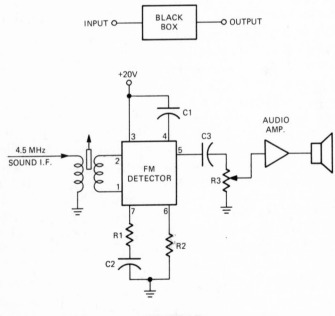

FIGURE 4-7
TYPICAL FM DETECTOR IC

89

diode probe of a millivolt meter. We can next look for audio across the volume control or at pin 5 of the FM-detector IC. If capacitor C3 were open, we would get audio at pin 5 but not at the volume control R3. If nothing appears at pin 5, then the reason for the defect may be in one of the external components, C1, C2, R1, or R2, or else it is in the FM detector itself.

In applying the systems concept approach to test the ICs, it is essential to know what the output must be for a given input. If, for example, we are checking the operation of an amplifier, we must know not merely the gain, but also the dynamic range of input signals over which the gain should be expected. If a control bias is supplied, we must know what its value is supposed to be. Much of this information is normally available from the equipment circuit diagram supplied by the manufacturer.

In troubleshooting digital logic, however, the situation is quite different. There are usually only two logic levels that can exist at any input or output. The manufacturer's handbook indicates, for each family of ICs, what the voltages should be for logic level 0 and for logic level 1. Once these values are known, it is possible to test logic gates or flip-flops simply by connecting ground or plus voltages to the appropriate input terminals and observing whether the output of the gate or flip-flop changes as it should for normal logic operation. This method may not work in CMOS or other high-speed logic where the relatively slow changes due to the application of clip leads will not change the state of a flip-flop.

Most digital ICs have a number of elements in a single package. A typical example, shown in Figure 4-8, contains four 2-input gates. It is possible to have one defective gate on a DIP unit while the other three gates work perfectly well. Once we have located such a defective gate, the question arises as to how to repair it. The proper way would be to replace the entire IC. Lazy troubleshooters have figured out that it is easier to run three wires to an empty gate on another IC on the same PC board than it is to unsolder the fourteen or sixteen pins of the DIP unit and solder a new one in place.

4.3 HOW ICs FAIL AND WHY

In chapter 3 we have discussed how transistors fail and why, and, since ICs are largely made up of transistors, resistors, and diodes, all

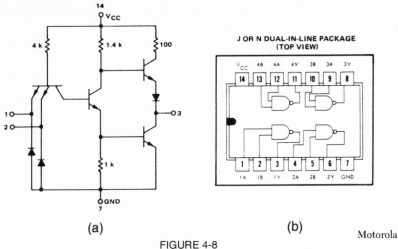

(a)

(b)

Motorola

FIGURE 4-8
TYPE 5400/7400 QUADRUPLE 2-INPUT NAND GATE

of the failure modes previously described for transistors will also apply to ICs. In addition, however, the extremely small size of the semiconductor elements and the substrate connections used in IC manufacture cause additional failure modes. The physical size of a typical transistor junction in a small-scale integrated circuit is approximately 1/25 the physical size of the transistor junction in a small-signal, low-power transistor. This means that any of the imperfections in manufacture that might cause a slightly deteriorated transistor performance in a single, discrete unit, will usually cause catastrophic failures in the IC. As one experienced troubleshooter put it so neatly, ICs don't get sick, they just die.

From our point of view, in troubleshooting electronic equipment the exact failure mode of an IC is not really important, particularly since we usually cannot tell whether there is a short or open circuit somewhere in a complex circuitry making up the IC. We can only tell whether the IC works or not. If we look at the circuit diagram of the operational amplifier in Figure 4-4, we have no way of knowing whether Q_6 or Q_5, Q_3, Q_8 or Q_9, etc., are either open or shorted. We only can tell that, with the proper signals applied at pin C and with the correct voltage at G and D, there is no output at F.

The most frequent cause of failure in digital ICs, those that generally operate in either the "on" or the "off," after they have been used in the equipment and operated for a while is usually an external cause

91

such as excessive voltage, excessive heat, or excessive mechanical stress. By mechanical stress we can think of such things as a damaged case, leads bent out of shape at the case entrance, cracks in the case insulation, etc. Excessive heat can be caused either by a failure within the IC causing excessive current and subsequently excessive temperature rise, or else by a temperature rise within the equipment due to the change in some other components or due to a defect somewhere else. Commercial-grade ICs are specified to operate up to $+70°$ C. Military-grade ICs are supposed to operate up to $+125°$ C.

ICs can also be damaged by being exposed to excessively high storage temperatures. For military-grade ICs the maximum storage temperature is $+150°$ C. and for commercial-grade ICs the maximum storage temperature is $+125°$ C. We can often tell when circuits have been exposed to temperature extremes because other components, including the printed circuit board and wires around it, will have been discolored.

IC damage due to excessive signals or power supply voltages is much harder to detect. We are always surprised to see how many equipment manufacturers do not provide overvoltage protection within their power supply circuits, knowing full well that a single transient, such as a 15- or 20-volt pulse riding on the 5-volt power supply, will destroy most of the standard 5-volt TTL circuits. MOS ICs, particularly some earlier versions, are also subject to static electricity building up between terminals, which can burn out the input semiconductors. As was explained in chapter 3 concerning the use of low-voltage soldering irons for transistors, it is absolutely essential to use only those soldering irons that are isolated from the AC power line by means of a step-down transformer. The leakage voltage that can develop between the AC power line and the soldering iron surface may well be sufficient to destroy the ICs you are working on.

4.4 HOW TO COMBINE SIGNAL TRACING WITH THE SYMPTOM-FUNCTION TECHNIQUE

We have seen from transistor circuits that the symptom-function technique is not very powerful when we want to isolate defects to within a single circuit. The symptom-function technique really helps only to isolate trouble to a particular circuit function. In many equipments a complete circuit function is performed by a single IC. When trouble is isolated to such a function we are then tempted to immediately replace that IC. In spite of what we have said above on IC failure modes, in general, ICs are far more reliable and less likely to fail than most of the other components that are used with them.

As a typical example of the type of problem for which the symptom-function technique must be combined with the signal-tracing technique, let us look at the color decoder circuit in Figure 4-9. This Motorola MC 1325 IC contains a total of ten transistors, a diode, and sixteen resistors, as shown in Figure 4-9a. Input to this circuit includes the color subcarrier at 3.58 MHz and the 3.58 MHz reference signal. The output of this circuit consists of the three color-difference signals, each at a low level, driving a transistor amplifier. Suppose that, using the symptom-function technique, we have determined that the lack of any color in the picture is due to the chroma demodulator. We have determined that the 3.58 MHz reference signal is properly generated and we have determined that the 3.58 MHz color subcarrier must also be available to the chroma demodulator. If we consider the circuitry in Figure 4-9a, the most likely defect of the overall chroma demodulator must be in the IC.

Before we remove the 14-pin DIP package from the PC board, however, a little signal tracing with the oscilloscope might help us avoid a lengthy unsoldering and resoldering job. Assume that we find the proper $+20$ volts at pin 14 of the IC and, checking carefully, we find that pin 7 is indeed grounded. Next we use the oscilloscope to check pin 13, pin 3, and pin 5 to make sure that the 3.58 MHz reference signal really appears at these pins. Similarly, we check pins 4, 6, and 8 to make sure that the chroma signal appears at these pins. From the detailed circuit in Figure 4-9a, we know that the three color-difference filters, each consisting of L_2, C_2, and C_3, go to the base of their respective output transistor. Clearly, if any of the C_2 or C_3 capacitors were shorted, the appropriate output would be lacking. The circuit in Figure 4-9b tells us that the signal at pins 2, 9, and 11 should appear to be a video signal, essentially the same as what we expect to see at the outputs on pins 1, 12, and 10. If signal tracing reveals that none of the color-difference video signals appear either across the filters or at the output, we should still not jump to the conclusion that the IC itself is defective. In Figure 4-9c, we have reproduced the maximum ratings of this device in order to show that, by exceeding these maximum ratings, it is possible to damage it and,

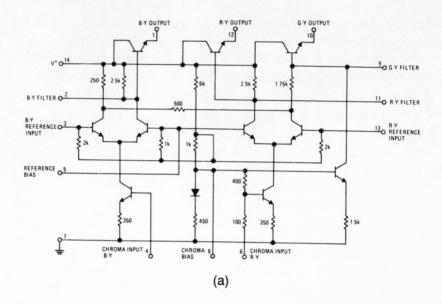

(a)

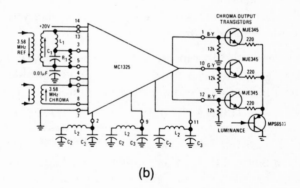

(b)

MAXIMUM RATINGS (T$_A$ = 25°C unless otherwise noted)

Rating	Symbol	Value	Unit
Power Supply Voltage	V$^+$	+24	Vdc
Reference Signal Input	e$_r$, e'$_r$	12	V$_{p-p}$
Output Current	I$_{out}$	30	mA
Chroma Signal Input	e$_c$	8.0	V$_{p-p}$
Power Dissipation (Package Limitation) Derate above T$_A$ = 25°C	P$_D$	625 5.0	mW mW/°C
Operating Temperature Range	T$_A$	0 to +70	°C

(c)

Motorola

FIGURE 4-9
CHROMA DEMODULATOR

if the unit is replaced, to damage the replacement as well. If, for example, the reference input at pins 3 or 13 were in excess of 12 volts peak to peak, or if the chroma signal input at pins 4 or 6 were in excess of 8 volts peak, we would expect to damage the input transistors. Measuring peak-to-peak voltages is relatively simple with the oscilloscope.

The TV receiver manufacturer's notes may tell us that the phase shift between the R-Y and the B-Y, pins 3 and 13, must be 105°, as determined by the values of C_1, L_1, and R_1. If any of these three components has changed substantially, the reference phases between R-Y and B-Y will be off, and correct color demodulation will not occur. In this case, however, incorrect colors, or some output at pins 1, 10, and 12, should be available. The signal-tracing method, as applied to the chroma demodulator in Figure 4-9, must be used with great care and understanding of the circuit. If, for example, any one of the three chroma output transistors in Figure 4-9b had a base-emitter short, this would be an effective short across the 12 K resistor and would cause excessive loading for the chroma demodulator output.

Because of the time-consuming task of unsoldering and resoldering ICs, it is always preferable to carefully check all other circuit elements, to check and recheck the input and output signals to the IC, and to compare the findings with the manufacturer's ratings. Combining the symptom-function technique with the signal-tracing technique is usually the most effective way of locating IC troubles. The only way we can usually be sure that we have located a defective IC is by replacing it with a new one and making sure that it works properly in the equipment.

4.5 LIMITATIONS OF THE VOLTAGE-RESISTANCE TECHNIQUE

The voltage-resistance technique of troubleshooting is useful in any kind of equipment where specific voltages and resistances are described in the manufacturer's data and where these parameters can easily be measured. In equipment predominantly using ICs, this technique is limited because, with the equipment turned on, we can essentially only be sure that the B+ voltages are correct and appear at the specified points. With the equipment turned off, we can only measure resistances where they actually exist in the circuit. Of course, short circuits that short out at the B+ can often be found by

95

this technique. If the short is caused by a defective IC, this may be apparent only when the power is turned on.

Measuring resistances between IC pins is usually not very useful because these resistances depend on the dynamic operation of a circuit within the IC package. For example, if we were to measure the resistance at the input of one of the gates of the IC illustrated in Figure 4-8, this resistance depends on the logic state, 0 or 1, of the preceding stage. In a dynamic digital circuit this condition will change according to the changes in the logic. In analog circuits, such as the op-amp illustrated in Figure 4-4, the input impedance will always be very low when the op-amp is operating. The output impedance will depend on the load the amplifier is driving. Neither of these values gives us any clue as to whether the IC is operating correctly or not.

In troubleshooting ICs, the signal-tracing and symptom-function techniques are so much more powerful than the voltage-resistance technique that the latter is used very rarely, and usually only to determine whether there is an obvious open or short circuit on the entire printed circuit board to which the defect has been isolated.

4.6 LIMITATIONS OF THE SUBSTITUTION TECHNIQUE

As we have seen in chapter 3, the substitution technique requires unsoldering in most cases. Unsoldering ICs is a tedious process and, unless extreme care is used, it is quite likely that the IC itself will be damaged in the process. For this reason, the substitution technique is rarely used in IC troubleshooting to locate a defective IC. Substitution is used only when the trouble has been traced down to a particular IC and substituting it with a known good one is the only method of confirming the diagnosis. In other words, we rarely use the substitution technique for troubleshooting but substitute ICs only as repair.

4.7 REPLACING ICs

When a new IC is soldered in place of the suspected one, we must be very careful to clean off all of the mounting holes of the IC before inserting the new one. Because the pins on every type of IC package are so close together, it happens very often that excess solder forms small bridges, and therefore short circuits between adjacent pins.

Many printed circuit boards use plated-through holes. This means that a conductor exists between the two sides of the IC board. In unsoldering the defective IC it can easily happen that this conducting sleeve is removed and the through-connection no longer exists. For this reason, experience troubleshooters always make sure that the substituted IC is properly soldered on both sides of the PC board so that, even if the plated-through hole has been damaged, the IC pin itself serves as conductor between the two sides of the printed circuit board.

If you have never unsoldered a 14- or 16-pin DIP package, the following procedure may help:

1. If a DIP unsoldering tool is available, use it with caution. This tool will heat all 14 or 16 lands on the PC board simultaneously. As soon as the solder flows, be prepared to remove the IC from the opposite side of the PC board with a pair of long-nosed pliers or a special IC-removal forceps. Don't plan to use your fingers since the PC board and the IC itself will be quite hot.

2. If you don't have an IC unsoldering tool, the best approach is to clip off the IC contacts close to the body of the package. Then unsolder each one individually and pull the contact gently out of its hole.

3. To remove excess solder, do not wipe the PC board since this may simply move the solder to other portions of the PC board and cause short circuits between adjacent conductors. The best method of removing excess solder is to use one of the Telflon-tipped suction bulbs available at electronics distributors.

4. Before inserting the new IC package, carefully inspect the area on the printed circuit board for excess solder, solder bridges, etc. Using an enlarging lens is very helpful at this point.

5. Carefully insert the new IC package, making sure that none of the pins are being bent as you insert them. Check the marking on the package to make sure that the pins are in the proper location.

6. With a very small soldering iron tip, carefully solder each pin in place, for double-sided PC boards solder both top and bottom.

7. Inspect your work, preferably with an enlarging glass, to make sure that all solder joints are clean and that excess solder does not form any bridges between adjacent pins or between adjacent conductors on the PC board.

If you had any doubts about the limitations of the substitution technique, the above description of an actual replacement job should

97

dispel these doubts. Never replace an IC unless absolutely sure that it is defective.

4.8 USING IC DATA TO ANALYZE POSSIBLE FAILURES

As a skilled electronics troubleshooter, you will be familiar with the use of manufacturer's data to determine whether a component is defective or not. In chapter 3 we have indicated the use of simple transistor characteristics to select transistors that might be substituted for another transistor. Understanding the manufacturer's data on any type of IC is only slightly more difficult than understanding transistor characteristics. Each of the major IC manufacturers publishes data books describing their products. When standard ICs are used, you will find the same essential description in any manufacturer's data book. The data sheet for the particular IC contains a wealth of information that, if properly used, can save you lots of time and trouble. Figures 4-10 and 4-11 show typical data sheets for a digital logic IC and Figure 4-12 shows a sheet describing the operation of an op-amp. The data that is important for digital logic, such as the maximum rise and fall time of the digital signals, is, of course, different from the data that is important for an analog circuit.

Without a detailed knowledge of the theory of digital circuits, we can quickly compare the signals, both the levels and the wave shapes, that are applied to the IC in the circuit and those that are specified by the data sheet. If we find that the input waveforms have a rise and fall time well within the capability of the circuit and then find that the output wave shapes are considerably distorted, we must assume that this is due either to the load of the circuit or due to a defect in the circuit itself. If we disconnect the load, this should eliminate that possible cause. In digital logic, we know that the output amplitude is not a direct function of the input amplitude. If the input signal is not large enough to turn the digital circuit on and off, as required, the output will not change. If the input signal, however, exceeds, even slightly, the minimum required, then the output should change, unless there is a short or an excessive load applied to the circuit. Disconnecting the load will again isolate the cause either to the load or to the circuit itself.

In analog circuits, as apparent from the data sheet, temperature and voltages are very important parameters. Here the output signal

ELECTRICAL CHARACTERISTICS

Test procedures are shown for only one inverter. The other inverters are tested in the same manner.

Inverters with pin pairs: 1→2, 3→4, 5→6, 7→8, 9→10, 11→12, 13.

Test Limits

Characteristic	Symbol	Pin Under Test	MC936, MC937 −55°C Min	Max	+25°C Min	Max	+125°C Min	Max	Unit	MC836, MC837 0°C Min	Max	+25°C Min	Max	+75°C Min	Max	Unit
Output Voltage	V_{OL}	2	−	0.40	−	0.40	−	0.45	Vdc	−	0.45	−	0.45	−	0.50	Vdc
	V_{OH}	2	2.50	−	2.60	−	2.50	−	Vdc	2.60	−	2.60	−	2.50	−	Vdc
Short-Circuit Current MC936, MC836	I_{SC}	2	−1.34	−4.00	−1.34	−4.00	−1.30	−3.90	mAdc	−1.30	−3.90	−1.30	−3.90	−1.25	−3.75	mAdc
MC937, MC837	I_{SC}	2	−1.34	−4.00	−1.34	−4.00	−1.30	−3.90	mAdc	−1.30	−3.90	−1.30	−3.90	−1.25	−3.75	mAdc
Reverse Current	I_R	1	−	2.0	−	2.0	−	5.0	µAdc	−	5.0	−	5.0	−	10	µAdc
Output Leakage Current	I_{CEX}	2	−	50	−	50	−	50	µAdc	−	100	−	100	−	100	µAdc
Forward Current	I_F	1	−	−1.60	−	−1.60	−	−1.50	mAdc	−	−1.40	−	−1.40	−	−1.33	mAdc
Power Drain Current (Total Device) MC936/MC836	I_{PDH}	14	−	−	19.5	−	−	−	mAdc	−	−	24	−	−	−	mAdc
MC937/MC837	I_{PDH}	14	−	−	32.0	−	−	−		−	−	39	−	−	−	
All Types	I_{max}	14	−	−	16.5	−	−	−		−	−	24	−	−	−	
Switching Times MC936, MC836	t_{pd+}	1,2	−	−	25	80	−	−	ns	−	−	−	80	−	−	ns
	t_{pd-}	1,2	−	−	10	30	−	−		−	−	−	30	−	−	
MC937, MC837	t_{pd+}	1,2	−	−	15	60	−	−		−	−	−	60	−	−	
	t_{pd-}	1,2	−	−	10	30	−	−		−	−	−	30	−	−	

Test Current / Voltage Values

MC936, MC937

@ Test Temperature	I_{OL} MC936 (mA)	I_{OL} MC937	I_{OH} MC936	I_{OH} MC937	V_{IL}	V_{IH}	V_F	V_R	V_{CEX}	V_{CC}	V_{CCL}	V_{CCH}	V_{max}	Gnd
−55°C	11.4	10.4	−0.12	−0.5	1.40	2.10	0	4.00	−	−	4.50	5.50	−	7
+25°C	12.0	11.0	−0.12	−0.5	1.10	2.00	0	4.00	4.50	5.00	4.50	5.50	8.00	7
+125°C	10.8	9.8	−0.12	−0.5	0.80	2.00	0	4.00	−	−	4.50	5.50	−	

MC836, MC837

@ Test Temperature	I_{OL} MC836 (mA)	I_{OL} MC837	I_{OH} MC836	I_{OH} MC837	V_{IL}	V_{IH}	V_F	V_R	V_{CEX}	V_{CC}	V_{CCL}	V_{CCH}	V_{max}	Gnd
0°C	12.0	11.0	−0.12	−0.5	1.20	2.00	0.45	4.00	−	−	5.00	5.00	−	7
+25°C	12.0	11.0	−0.12	−0.5	1.10	1.90	0.45	5.00	5.00	5.00	5.00	5.00	8.00	7
+75°C	11.4	10.4	−0.12	−0.5	0.95	1.80	0.50	4.00	−	−	5.00	5.00	−	

Test Current / Voltage Applied to Pins Listed Below

Characteristic	I_{OL}	I_{OH}	V_F	V_R	V_{CEX}	V_{CC}	Pulse In	Pulse Out	Gnd
Output Voltage	2	2							7
Short-Circuit Current		2							1,2,7
Reverse Current				1					7
Output Leakage Current					2,14				1,7
Forward Current			1						7
Power Drain MC936/MC836						14			7
Power Drain MC937/MC837						14			7
All Types									1,3,5,7,9,11,13
Switching Times						14	1	2	7

Pins not listed are left open.

FIGURE 4-10
DIGITAL IC CHARACTERISTICS FOR MC936, ETC.

Motorola

SWITCHING TIME TEST CIRCUIT AND WAVEFORMS

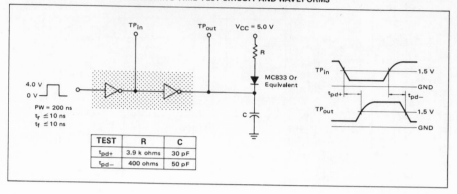

FIGURE 4-11   Motorola
TEST CIRCUIT FOR DATA IN FIGURE 4-10

should follow the input signal closely. If no input signal is apparent at an op-amp, we should not expect an output. External component values, as indicated by the data in Figure 4-12, can be extremely critical. It is often possible to test an analog IC by using external components, which can easily be soldered temporarily into the circuit according to the values given in the data sheet. If the relationship of input and output is then within the specified limits, the IC is good and the fault must lie elsewhere. In the case of the narrow-band amplifier in Figure 4-5, for example, the data sheet may not show the feedback capacitor, C_1, but instead a different value for R_3. The connections to pin 6 may be given as a short circuit to ground and the connection to pin 3 may also be a different resistor-capacitor network. It is often simpler to change these external parts to conform to the test conditions described in the data sheet and then vary the input signal, measure the output signal, and see whether the amplifier conforms to the specifications. This is a method of testing the IC in the circuit, simply by temporarily disconnecting the existing external components and connecting new ones in their place. It is much easier to do this than to replace 14- or 16-pin ICs just on the suspicion that they might be defective.

4.9 TYPICAL IC TROUBLES

As we explained in paragraph 4.3 above, ICs usually fail by being either completely out of specification or else simply "dead." The vast

ELECTRICAL CHARACTERISTICS (T_A = 25°C unless otherwise noted)

Characteristic Definitions ①	Characteristic	Symbol	Min	Typ	Max	Unit
$A_{VOL} = \dfrac{e_{out}}{e_{in}}$ (diagram with e_{in}, Z_{in}, Z_{out}, e_{out}, C, G, B)	Open Loop Voltage Gain R_L = 100 kΩ	A_{VOL}				V/V
	(V^+ = 6.0 Vdc, V^- = -3.0 Vdc, (V_{out} = ± 2.5 V)		600	900	1500	
	(V^+ = 12 Vdc, V^- = -6.0 Vdc, V_{out} = ± 5.0 V)		2500	3600	6000	
	(V^+ = 12 Vdc, V^- = -6.0 Vdc, V_{out} = ± 5.0 V, T_A = -55, +125°C)		2000		7000	
	(V^+ = 6.0 Vdc, V^- = -3.0 Vdc, V_{out} = ± 2.5 V, T_A = -55 to +125°C)		500		1750	
	Output Impedance	Z_{out}				ohms
	(V^+ = 6.0 Vdc, V^- = -3.0 Vdc, f = 20 Hz)		-	300	700	
	(V^+ = 12 Vdc, V^- = -6.0 Vdc, f = 20 Hz)		-	200	500	
	Input Impedance	Z_{in}				k ohms
	(V^+ = 6.0 Vdc, V^- = -3.0 Vdc, f = 20 Hz)		22	70	-	
	(V^+ = 6.0 Vdc, V^- = -3.0 Vdc, f = 20 Hz, T_A = -55°C, +125°C)		8.0	-	-	
	(V^+ = 12 Vdc, V^- = -6.0 Vdc, f = 20 Hz)		16	40	-	
	(V^+ = 12 Vdc, V^- = -6.0 Vdc, f = 20 Hz, T_A = -55°C, +125°C)		6.0			
(diagram, C, G, B, 0 V, +V, -V)	Output Voltage Swing	V_{out}				V_{peak}
	(V^+ = 6.0 Vdc, V^- = -3.0 Vdc, R_L = 100 kΩ)		± 2.5	± 2.7	-	
	(V^+ = 12 Vdc, V^- = -6.0 Vdc, R_L = 100 kΩ)		± 5.0	± 5.3	-	
	(V^+ = +6.0 Vdc, V^- = -3.0 Vdc, R_L = 10 kΩ)		± 1.5	± 2.0		
	(V^+ = +12 Vdc, V^- = -6.0 Vdc, R_L = 10 kΩ)		± 3.5	± 4.0		
$A_{VCM} = \dfrac{e_{out}}{e_{in}}$ $CM_{rej} = A_{VCM} - A_{VOL}$ (diagram, e_{in}, C, G, B, e_{out})	Input Common Mode Voltage Swing	CMV_{in}				V_{peak}
	(V^+ = 6.0 Vdc, V^- = -3.0 Vdc)		+0.5 -1.5	- -	- -	
	(V^+ = 12 Vdc, V^- = -6.0 Vdc)		+0.5 -4.0	- -	- -	
	Common Mode Rejection Ratio	CM_{rej}				dB
	(V^+ = 6.0 Vdc, V^- = -3.0 Vdc, f ≤ 1.0 kHz)		80	100	-	
	(V^+ = 12 Vdc, V^- = -6.0 Vdc, f ≤ 1.0 kHz)		80	100	-	
(diagram, I_2, C, G, B, I_1)	Input Bias Current T_A = 25°C	I_b				µA
	$I_b = \dfrac{I_1 + I_2}{2}$, (V^+ = 6.0 Vdc, V^- = -3.0 Vdc)		-	1.2	3.5	
	(V^+ = 12 Vdc, V^- = -6.0 Vdc)		-	2.0	5.0	
	T_A = -55°C (V^+ = 6.0 Vdc, V^- = -3.0 Vdc)		-	2.5	7.5	
	(V^+ = 12 Vdc, V^- = -6.0 Vdc)		-	4.0	10	
(diagram, I_2, C, G, B, I_1)	Input Offset Current ($I_{io} = I_1 - I_2$)	I_{io}				µA
	(V^+ = 6.0 Vdc, V^- = -3.0 Vdc)		-	0.1	0.5	
	(V^+ = 6.0 Vdc, V^- = -3.0 Vdc, T_A = -55 to +125°C)		-	-	1.5	
	(V^+ = 12 Vdc, V^- = -6.0 Vdc)		-	0.2	0.5	
	(V^+ = 12 Vdc, V^- = -6.0 Vdc, T_A = -55 to +125°C)		-	-	1.5	
(diagram, C, G, B, V_{in}, V_{out} = 0)	Input Offset Voltage R_S = 2.0 kΩ	V_{io}				mV
	(V^+ = 6.0 Vdc, V^- = -3.0 Vdc)		-	1.3	3.0	
	(V^+ = 6.0 Vdc, V^- = -3.0 Vdc, T_A = -55°C, +125°C)		-	-	4.0	
	(V^+ = 12 Vdc, V^- = -6.0 Vdc)		-	1.1	2.0	
	(V^+ = 12 Vdc, V^- = -6.0 Vdc, T_A = -55°C, +125°C)		-	-	3.0	
(step response diagram with e_{in}, 50%, 10%, 90%, e_{out}, SLEW RATE, R_1, R_2, R_3, C_1, C_2, R_L, C_L, R_L = 100 kΩ, C_L ≤ 100 pF)	Step Response V^+ = 12 Vdc, V^- = -6.0 Vdc, Gain = 100, V_{in} = 1.0 mV, R_1 = 1.0 kΩ, R_2 = 100 kΩ, C_2 = 50 pF, R_3 = ∞, C_1 = open	V_{os} t_r t_{pd} dV_{out}/dt②	- - - -	20 10 10 12	40 30 - -	% ns ns V/µs
	V^+ = 12 Vdc, V^- = -6.0 Vdc, Gain = 1.0, V_{in} = 10 mV, R_1 = 10 kΩ, R_2 = 10 kΩ, C_1 = 0.01 µF, R_3 = 20Ω, C_2 = open	V_{os} t_f t_{pd} dV_{out}/dt②	- - - -	10 25 16 1.5	50 120 - -	% ns ns V/µs
	Average Temperature Coefficient of Input Offset Voltage R_S = 50 Ω	TC_{Vio}				µV/°C
	(T_A = +25 to +125°C)		-	2.5	-	
	(T_A = -55 to +25°C)		-	2.0	-	
	Average Temperature Coefficient Input Offset Current	TC_{Iio}				nA/°C
	(T_A = +25°C to +125°C)		-	0.05	-	
	(T_A = -55 to +25°C)		-	1.5	-	
	DC Power Dissipation	P_D				mW
	(V_{out} = 0, V^+ = 6.0 Vdc, V^- = -3.0 Vdc)		-	17	30	
	(V_{out} = 0, V^+ = 12 Vdc, V^- = -6.0 Vdc)		-	70	120	
(diagram, C, G, B, V^+, V^-, SENSITIVITY = S, V_{out}, $S = \dfrac{\Delta V_{out}}{\Delta V_S (A_{VOL})}$)	Positive Supply Sensitivity (V^- constant = -6.0 Vdc, V^+ = 12 Vdc to 6.0 Vdc)	S^+	-	60	200	µV/V
	Negative Supply Sensitivity (V^+ constant = 12 Vdc, V^- = -6.0 Vdc to -3.0 Vdc)	S^-	-	60	200	µV/V

① All definitions imply linear operation. ② dV_{out}/dt = Slew Rate

Motorola

FIGURE 4-12
ANALOG IC CHARACTERISTICS FOR MC1712 OP-AMP

majority of IC failures, as far as the troubleshooter is concerned, is indicated by the lack of an output when a proper input is applied. This holds particularly true for digital ICs, but analog ICs, as well, are most likely to fail totally. The reason for this is that most IC failures are due to a mechanical defect, causing either open or short circuits.

Occasionally, the defect involves an internal failure that does not result in a completely "dead" unit. A typical example of this type of defect occurred during the troubleshooting of a mini-computer. By using the symptom-function technique, the defect was isolated to the basic circuitry shown in Figure 4-13. In this circuit a 180 KHz clock-signal is counted down, in two successive IC stages, to 3 KHz. The 3 KHz signal forms a gate for the 180 KHz clock pulse, allowing 60 pulses to pass through the AND gate during alternate 3 KHz periods, as shown in Figure 4-13b. When connecting the scope to point C, however, the waveform observed was twice the rate as illustrated. At point D, the other input to the AND circuit, the expected 180 KHz continuous signal was present. Clearly, then, something was wrong in

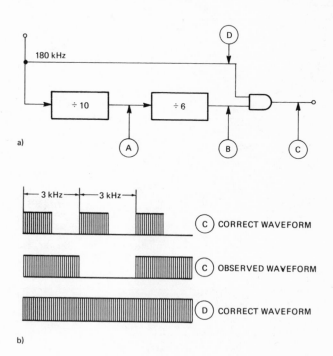

FIGURE 4-13
EXAMPLE OF DIGITAL IC TROUBLESHOOTING

the two ICs providing the division by 10 and by 6. Connecting the scope to point B, we found a 6 KHz square wave, twice the frequency expected. Moving to point A, the oscilloscope showed a 36 KHz square wave where we would have expected to find an 18 KHz square wave. Clearly then, the defect was in the "divide-by-10" IC. Replacing this IC with a new one did not clear up the problem. Why should an IC designed to divide by 10 suddenly decide to divide only by 5? The mystery was finally solved when we looked at the manufacturer's connection diagram, shown in Figure 4-14, for the "divide-by-10" IC. It became apparent that the output of the portion dividing by 5 was shorted to the subsequent "divide-by-2" stage. Careful examination of the PC board revealed that a small solder bridge had formed across the conductors connecting pin 11 and pin 12, shorting the "divide-by-2" stage (pin 14 in, pin 12 out). Clearing up this short circuit and soldering the old IC in place again revealed that the IC itself had not been bad to begin with.

In another typical IC troubleshooting example in a mini-computer, it was found that an LSI IC, used for converting BCD input to straight decimal output, was obviously defective, since no output could be obtained. Replacing it with another IC did not cure the problem. A third IC was temporarily soldered in place and it, too, failed to operate. A careful inspection of all of the lands of the PC board revealed no particular short or open circuit. One of the three suspected ICs was then soldered into another PC board on another mini-computer,

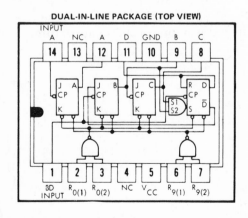

Texas Instruments

FIGURE 4-14
CONNECTIONS OF 7490 DECADE COUNTER

where it also appeared defective. The other two ICs were next tried in this "test position" and were also found to be defective. Clearly, something was wrong in the original PC board, ruining the ICs as they were soldered in place. We finally unplugged the PC card and plugged one proven to be good in the "test position" in its place. Promptly the same IC also became defective. We next searched through the wiring of the PC socket and found, eventually, that every time the mini-computer was turned on, a large transient pulse, apparently in the order of 30 or 40 volts, was applied through the PC board connector to the clock input of the particular IC. Naturally, this high voltage would blow every IC that was put into this position. The eventually located culprit was an open Zener diode that the manufacturer had carefully placed elsewhere in the system to prevent just this type of defect.

A typical analog IC problem was one in which the operational amplifier oscillated, no matter what adjustments were made in the feedback resistance. It was eventually found that an open circuit must have occurred in the pin connecting the feedback resistor into the amplifier. This open circuit apparently existed just inside of the IC package. Replacing the package with a good one cured the defect.

The typical IC troubles discussed here were selected from the combined experience of a number of troubleshooters in order to illustrate that we cannot simply, every time, assume that the IC is defective and replace it. It is true that in the majority of cases you will encounter a defective IC as the source of the problem, whether the defect is a short, an open or an intermittent one, or whether the IC somehow has drifted way out of the manufacturer's specifications.

Troubleshooting
Electron Tube Circuits

5.1 A BRIEF REVIEW OF TUBE FUNDAMENTALS

When we mention tubes, most of us think of vacuum tubes, although, to be accurate, all types of electron tubes, including cathode ray tubes, camera tubes, and gas tubes, should be included in this term. This chapter is devoted to all kinds of electron tubes but, since the majority of troubleshooting tasks still involve vacuum tubes, emphasis will be placed on the types of vacuum tubes most frequently found in electronic equipment.

The simplest vacuum tube uses only two elements, a heated filament, which acts as cathode, and the anode or plate. Electrons in a vacuum will travel from a heated surface to another one charged positively. In the illustration in Figure 5-1, the basic principles of vacuum and gas diodes are briefly illustrated. As shown, an AC voltage is supplied, through a transformer, to the filament, which acts as cathode and which emits electrons. The filament could just as well be heated by DC, but AC is usually used. The anode or plate is positively charged through the battery and will attract the electrons that are emitted from the cathode. The conventional designation of electron flow is opposite to that of current flow so that the current I is considered to be flowing from the plate to the cathode. The relationship between the voltage E and the current I is shown below the diode circuit. When zero voltage is applied, zero current flows. As the voltage is increased, more current will flow until a level is reached beyond which an increase in voltage produces only a very small or negligible increase in current. This is due to the fact that the cathode is limited in the amount of electrons it can emit. If a negative voltage is applied to the plate, with respect to the cathode, very little current will flow in that direction.

105

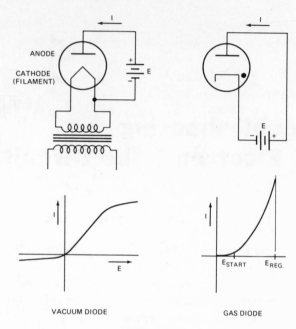

VACUUM DIODE   GAS DIODE

FIGURE 5-1
DIODE PRINCIPLES

In vacuum tubes it is essential to heat the cathode in order to get the emission of electrons. When the inside of the tube envelope is filled with certain gases, however, it is possible to get current flowing from the cathode to the anode without heating up the cathode. These so-called "cold-cathode" tubes are frequently used as rectifiers in industrial applications or as glow tubes, like the familiar neon tube. There are also some specialized types of gas tubes that do have heated filaments.

As illustrated in Figure 5-1, the current-voltage characteristic of a gas diode is quite different from that of a vacuum diode. Until the minimum starting voltage E_S has been reached, no current flows through the tube at all. This starting voltage is also called the ionizing potential and is dependent on the type of gas, the density of the gas, and the spacing between cathode and anode. At the ionizing potential the gas molecules become ionized, a glow is visible and current flows from the plate to the cathode. As the voltage E is increased, the current increases rapidly. At the point where even a very small increase in voltage causes a very large increase in current, the gas diode

106

is considered to be regulating. In many industrial electronic devices this property is used to provide voltage regulation since gas tubes are capable of varying their current very widely with very small variations in voltage. If the applied voltage is increased beyond the regulating voltage and reaches the limitation of the current-carrying capability of the gas tube, breakdown and arcing occurs, which will damage the tube.

Diodes are two terminal devices and are primarily used for rectification, as switches, and, in the case of the gas diode, for voltage regulation and to emit light. The triode is the basic vacuum tube used for amplification of signals. Figure 5-2 illustrates the DC operating prin-

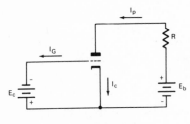

FIGURE 5-2
DC TRIODE PRINCIPLES

ciples of the basic triode. As in the case of the diode, a plate (anode) and a cathode are used, with current flowing from the plate to the cathode. Between the plate and cathode, closest to the cathode, is a wire grid or mesh that is set at a different potential. Electrons leaving the cathode and traveling toward the plate can pass through this wire mesh. If, however, this wire mesh is placed at a potential negative with respect to the cathode, many of these electrons will be deflected back to the cathode. If the grid is placed at the same potential as the cathode, it will have no effect. If the grid is positive with respect to the cathode, it attracts many of the electrons traveling to the plate and grid current flows. For class A amplification, the control grid, as it is usually called, is always negative with respect to the cathode. The signal to be amplified changes the degree of negative voltage and therefore controls the amount of electrons flowing from the cathode to the anode. Stated in more practical terms, the plate current is controlled by the grid voltage. Until the control grid reaches a positive voltage, with respect to the cathode, it draws practically no current.

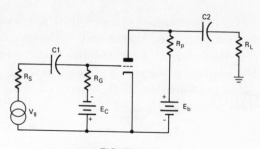

FIGURE 5-3
BASIC TRIODE AMPLIFIER

A basic triode amplifier is shown in Figure 5-3. Although separate batteries are shown for E_b and E_C, the plate voltage and grid bias, respectively, these DC voltages are obtained in a different way in actual circuits. The AC signal, represented by the generator V_g, and its source resistance R_S is applied to the control grid through coupling capacitor C1 and R_G, which represents the load resistance for the signal generator. The bias provided by E_C is usually set well below the point at which R_G would draw current through the grid circuit itself. The output signal is the voltage developed across R_p, due to variations in plate current. This voltage is coupled through C2 to the final load, R_L.

An actual triode circuit is shown in Figure 5-4. Note that the grid bias battery E_C is replaced by the cathode resistor and capacitor combination, R_K and C_K. With plate current flowing through R_p and R_K, a voltage is set up across R_K making the cathode more positive than the

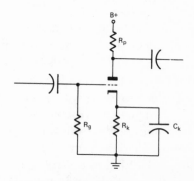

FIGURE 5-4
CATHODE SELF-BIAS CIRCUIT

108

ground. Since the grid resistor R_G is returned to ground, this makes the control grid negative with respect to the cathode and provides the same effect as E_C in Figure 5-3. The capacitor C_K is necessary to provide an effective short circuit across R_K for the AC signals. Without C_K the AC output signals appearing across R_p would be reduced due to the degenerative effect of R_K.

The operation of a triode can be simply described by means of the three sets of characteristic curves shown in Figure 5-5 a, b, c. The plate characteristics show the relationship between the plate voltage e_b and the plate current i_b. Different, but essentially parallel, curves will occur for different values of grid voltage. With the grid voltage varying, as a result of a sinewave input signal, for example, the plate current itself will vary if the plate supply voltage is kept fixed. The slope of these curves is the inverse of the plate resistance, the effective tube resistance, not the external plate load resistance. The second curve, essentially a straight line, represents the constant current characteristics. Here the variations of the grid voltage ec and the plate voltage eb are shown for fixed values of plate current. As grid voltage ec is varied and the plate voltage eb is kept constant, the changes in plate current can be observed. Note that the slope of the plate current curves is $-\mu$, where μ is called the amplification factor of a tube. The third set of characteristics shown in Figure 5-5 is the transfer characteristics, which illustrate the relationship between variations in grid voltage with fixed plate current. The slope of these curves is the gm, the transconductance of the tube. When analyzing vacuum tube performance, the basic relationship between the three characteristics can be obtained from the following equation:

$$\mu + rp\ gm$$

As in the case of transistors, triodes can be connected in three types of circuitry. Figure 5-6 illustrates the three triode connections, together with their respective equivalent circuits. The basic triode amplifier illustrated in Figures 5-3 and 5-4 corresponds to the grounded cathode configuration of Figure 5-6a. In this arrangement the input impedance is quite high and the output impedance is essentially a function of the vacuum tube's own plate resistance r_p. In the grounded grid configuration, which corresponds essentially to the grounded base configuration in transistors, the input impedance is relatively low, depending on the cathode circuit, and the output impedance depends, as before, on the plate circuit impedance. The

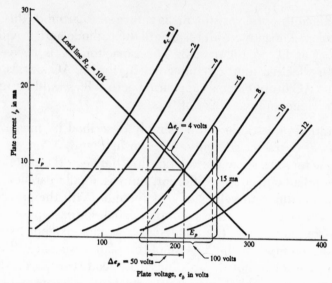

Lane K. Branson, *Introduction to Electronics*, © 1967. Reprinted by permission of Prentice-Hall, Inc.

a) Plate characteristics

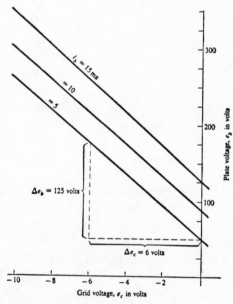

Lane K. Branson, *Introduction to Electronics*, © 1967. Reprinted by permission of Prentice-Hall, Inc

b) Constant current characteristics

FIGURE 5-5
TRIODE CHARACTERISTICS

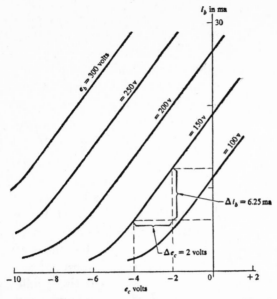

Lane K. Branson, *Introduction to Electronics*, © 1967. Reprinted by permission of Prentice-Hall, Inc.

c) Transfer characteristics

FIGURE 5-5 *CONT.*

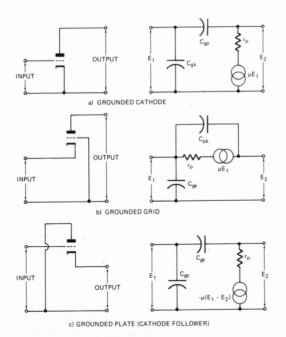

a) GROUNDED CATHODE

b) GROUNDED GRID

c) GROUNDED PLATE (CATHODE FOLLOWER)

FIGURE 5-6
THREE POSSIBLE TRIODE CONNECTIONS

111

third configuration, grounded plate, is generally called the cathode follower and has some unique characteristics. Because the plate is essentially grounded, at least for the signal frequencies, the output appears across the cathode load. This makes it possible to have a very high input impedance and a very low output impedance, essentially the cathode impedance. Voltage gain is not possible in this circuit with the theoretical maximum gain equal to 1. Power gain, however, is possible because of the change in input and output impedance.

In addition to triodes, a variety of other tubes that provide special operating performance are on the market. In Figure 5-7, we have illustrated a tetrode circuit and a pentode circuit as being the most widely used types of multigrid tubes. In the tetrode circuit a second grid, called the screen grid, is placed between the control grid and the anode, generally closer to the anode than the control grid. It helps the plate attract the electrons and it is therefore placed at a positive potential generally somewhat less than the anode potential. The screen grid also provides a constant current characteristic since, when the plate voltage itself is low due to the current swing across the plate load resistor, the screen grid will still attract the same amount of electrons from the cathode.

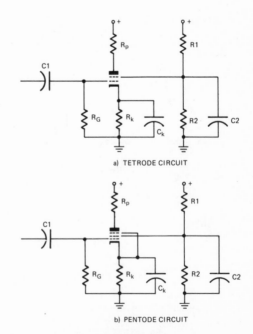

a) TETRODE CIRCUIT

b) PENTODE CIRCUIT

FIGURE 5-7
TYPICAL MULTIGRID TUBE CIRCUITS

In the pentode circuit, a third grid has been added and this grid is placed close to the anode structure itself. By placing this third grid at the same potential as the cathode it helps to turn back electrons that have been knocked loose from the anode as a result of the impact of electrons speeded up to the anode by the screen grid. In many circuits this third or suppressor grid is used to apply a second input signal, in addition to the one that appears on the control grid.

5.2 TUBE TESTING AND SUBSTITUTION

WATCH OUT: TUBES GET VERY HOT AND BURNED FINGERS ARE NOT A SIGN OF THE COMPETENT TROUBLESHOOTER. HIGH VOLTAGES ARE USED IN TUBE CIRCUITS—DO NOT TOUCH EXPOSED WIRES OR TEST POINTS.

In all receiving-type tubes the type number starts out with the filament voltage. Vacuum tubes starting with 6, followed by letters, such as 6CB6, require 6.3 volts across the filament. The precision of this voltage depends on the tube type and application, but in general, a 6-volt filament will operate satisfactorily between 5.8 and 6.5 volts. Almost invariably, 6-volt filament tubes are used in parallel arrangements as shown in Figure 5-8. In this type of circuit the same voltage is applied to all tubes but the current through each filament depends on the filament's resistance. In a typical, old-style TV receiver using parallel filaments, some tubes will heat up faster than others due to the amount of current drawn by their filaments.

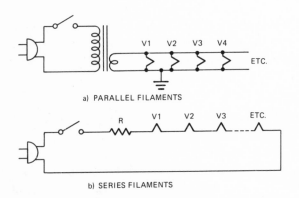

a) PARALLEL FILAMENTS

b) SERIES FILAMENTS

FIGURE 5-8
BASIC FILAMENT CIRCUITS

113

In later TV receiver models, and in a host of broadcast radios and FM sets, the series filament approach was used, as illustrated in Figure 5-8. In this circuit the same current flows through every filament but the voltage across it differs. A 35W4 vacuum tube, a favorite in many series filament radio receivers, requires 35 volts across its filament. It is obvious from the circuit in Figure 5-8 that the total voltage across all of the filaments connected in series must be less than 117 volts. The series resistor shown serves to limit the surge current through the filaments. In many applications this resistor has a high resistance when cold and a relatively low resistance when hot.

In electronic equipment using vacuum tubes the most likely component to become defective is the tube itself. As will be explained in Section 5.3, the most likely way for a vacuum tube to fail is by an open filament. Since filament defects are the most frequent cause for tube failure, one simple way of checking tubes is by looking at them and observing whether the filament glows or not. Metal tubes and special-purpose tubes in which the filament is not visible can be checked cautiously by touching with a wet finger. This method works only when parallel filaments are used because, in a series filament arrangement, the failure of a single tube filament will prevent the current from reaching any of them. Fortunately, it is relatively simple, with the AC plug removed from the socket, to check a series filament string for continuity with the ohmmeter. When the ohmmeter reading shows an open circuit, you know that this tube is defective.

If the tube lights up, this is an indication that the filament is working correctly and some other fault must cause the tube failure. Tube testers are in very widespread use and a variety of inexpensive, handy, and portable tube testers are available to the electronics troubleshooter. Neglecting the drugstore variety of tube tester, we need to know only a few things in order to be able to test tubes on our tube tester.

The variety of different tube sockets require that the tube tester have sockets to match each type of tube. Before we plug the suspected tube into its correct socket, we should set the tube tester controls for the particular tube type. Most testers have internal roll-charts in which we can look up the tube type and learn how to set the various switches and controls for that particular tube. These controls set the operating conditions, such as filament voltage, grid bias, plate voltage, screen voltage, etc., and complete the test circuit. After the tube-tester controls have been set, we plug in the tube and then simply follow the instructions, which are usually printed on the front panel. These instructions invariably require us to check first whether a short circuit exists between tube elements. Plate and screen voltage is applied by depressing the "test" button and then we can see whether the specified plate current is drawn and whether the specified transconductance gm or amplification factor is being measured.

In essence, the tube tester measures some of the tube characteristics described by the graphs in Figure 5-5. It does not measure any of the interelectrode capacities, as illustrated in the equivalent circuits of Figure 5-6, and it does not measure the dynamic performance of the tube, such as linearity in class A operation or power output in class C operation.

Whenever the results of the tube tests show the slightest suspicion that the tube might be marginal or defective, it is obviously a good idea to replace the tube with a new one. One of the problems with tube testing, however, is that tubes that appear to be operating perfectly on the tube tester will sometimes fail in the circuit. This is particularly true of circuits in which the interelectrode capacities are critical or circuits in which the full dynamic tube characteristic curves are important. A vacuum tube may work perfectly fine as an amplifier, but when it is used as a limiter, such as in FM receivers, the plate voltage-plate current curves may not be sufficiently flat to provide good limiting. In oscillators, such as the horizontal sweep circuit in a TV receiver or in a UHF tuner, a tube that checks perfectly good on the tube tester may produce either the wrong wave shape or may not oscillate at all. Tubes that are required to operate at high voltage or in high peak-current applications may also show healthy on the tube tester but may not perform in the circuit.

If we cannot rely on the tube tester to prove that a tube is defective or good, why use it? Many troubleshooters, particularly those who work mostly with solid-state devices, don't bother with a tube tester at all. If a tube is suspected, they figure it is cheaper to try a new tube than to bother testing the suspected one on the tube tester and, chances are, having to replace it with a new one anyway.

In chapters 3 and 4 we have discussed the substitution technique for transistors and for integrated circuits. We have found that substitu-

115

tion is not a good way of troubleshooting transistors and integrated circuits. For vacuum tubes, however, substitution is the very best, simplest, and easiest way of finding the most likely defect, the vacuum tube. Vacuum tubes can be substituted very easily and the only limitation is that we must have the exact replacement tube or its exact replacement equivalent.

Before you start substituting tubes, however, we must give you this one word of caution:

TURN THE EQUIPMENT OFF BEFORE REMOVING THE OLD TUBE AND SUBSTITUTING THE NEW ONE.

The temptation is always great to pull out the suspected tube, particularly if it is cold and obviously defective, and simply plug in the new tube. In some circuits this will not harm anything, but in general it is bad practice. When the tube is plugged in, even in parallel filament circuits, it takes a few seconds for the filament to heat up, but the plate voltage is already applied at that time, something that may damage certain types of tubes. The only exception to this rule should be made in the case of tube failure that shows only after the entire equipment has been on and is hot. Then it may be helpful in your troubleshooting if you can quickly yank out the suspected tube and plug in a new, good one. If the defect now disappears, it's clearly the tube. If the defect is still there, even with the new tube, then some other component is getting hot and changing its value beyond circuit tolerances.

Substitution of tubes is a simple and usually effective way of testing tubes and the only limitation to this approach is the need for keeping a stock of replacement tubes on hand.

5.3 HOW TUBES FAIL AND WHY

The layman does not care why a particular tube has failed or what kind of defect has occurred as long as a replacement tube is handy, can be plugged in and the equipment works properly again. The professional electronics troubleshooter, however, needs to know how tubes fail and why they fail, so that he can avoid repeated tube failures.

In the case of transistors we have learned that a shorted diode junction is the most frequent failure mode, but in vacuum tubes the most frequent failure mode is the open filament. Before the availability of transistors, when vacuum tubes were used in every kind of equipment, extensive studies were made as to the reasons for open filaments. It was found that filaments were weakened by frequent current surges. The heating and cooling cycle, inherent in turning equipment on and off, tends to shorten the life of filaments. Mechanical vibration and shock also contribute to filament failure and special tube designs were developed with extra-rugged filaments. A third, occasional, type of filament failure was due to a relatively high voltage applied between the filaments and the cathode. To overcome this problem, filaments were either well insulated from the cathode or else the external circuitry was arranged to avoid this potential difference between the two elements.

Most receiving tubes, particularly those designed for consumer equipment, can be mounted in any position. High-power tubes, such as transmitting tubes, high-power rectifiers, or special-purpose tubes, however, may have to be mounted in a vertical position to reduce the effect of gravity on the filament, the cathode, or the grid elements themselves. This is particularly important in large vacuum tubes where the length of the internal structure is greater than half an inch or so. Sagging filament or grid wires, which become soft due to the heat, break sooner and have a tendency to short out other elements. Ordinarily, the original equipment design accounts for the proper mounting of the vacuum tube but we have occasionally found that equipment that was meant to "stand up" has been "laid down" by the user and this has resulted in a shortened tube life of expensive high-power vacuum tubes.

Quite unlike transistors, which can theoretically work forever, vacuum tubes have a limited life because the emissive coating on the cathode will eventually erode and the tube will become weaker, i.e., fewer and fewer electrons can be emitted from the cathode. This type of defect shows easily on any kind of tube tester because the meter will simply indicate a low transconductance by reading "weak."

When a slight amount of gas is generated within the vacuum tube envelope, the tube's operation is greatly impaired and we generally call this a gassy tube. The exact reason for this type of defect is difficult to determine but is usually due to chemical reactions among the various vacuum tube elements. If you have ever looked at the inside of a vacuum tube, you will know that the internal structure contains a number of fine wires, small, thin parts, etc., all welded in a carefully aligned assembly. It is possible, due to mechanical vibration and shock, or simply due to faulty construction, that some little metal piece may break off. This then results in floating debris, which can cause intermittent or permanent shorts among the elements within the glass envelope. Depending on the location of this stray piece of metal, the tube may be intermittently shorted, elements may be shorted together, occasional arcing may occur, or other problems may arise.

The chance that wrong voltages damage a vacuum tube is much less than in the case of transistors but it can still occur occasionally. If you short the plate and grid together for any length of time, the resulting grid current will tend to damage the fine wires of the grid. Because a positive grid causes a lot of current to be drawn from the cathode, it will probably damage the cathode structure in a few minutes. Similarly, some tubes can be damaged when they are operated without the proper grid bias because the screen grid and plate structure may not be capable of withstanding the high heat generated by the excess current. On rare occasions excessive voltage can be applied to a screen or cathode, resulting in internal arc-over. This situation is more likely in circuits using voltages in excess of 500 volts. While vacuum tubes normally operate quite hot and they are therefore not as temperature-sensitive as semiconductors, excess heat will accelerate the life span of a vacuum tube and will cause it to burn out much faster. In chapter 6, Finding Intermittent Defects, we will show you a method of smoking out intermittent vacuum tube defects involving the closing-up of the equipment, allowing heat to build up, and thereby converting intermittent defects into permanent ones.

In addition to internal problems, vacuum tube failure is often caused by poor connections outside the glass envelope. Tube sockets are very handy since they permit us to rapidly change suspected tubes, but they are also a source of trouble in themselves. Whenever a tube pin makes poor contact with the socket, trouble can be expected. In many tube types a stiff wire passing through the glass seal of the tube envelope is the means of making contact. In octal tubes and in most types of cathode ray tubes, however, a different arrangement is used, as illustrated in Figure 5-9. The wire passing through the glass seal goes through the phenolic tube base into a hollow contact and it is this contact sleeve itself that plugs into the socket. At the factory, the wire

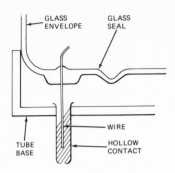

FIGURE 5-9
TUBE BASE CROSS SECTION

is soldered or welded to the hollow contact. We have seen many instances where this wire has come loose from the contact sleeve and as a result, no connection or an intermittent connection is made to the circuitry. The shaded portion of Figure 5-9 indicates where solder fill should be. If this type of defect is suspected, it is only necessary to apply a hot soldering iron to the outside of the hollow contact to make sure that the solder inside gets reheated and makes good contact. The bottom of this hollow contact is open and it is often possible for the troubleshooter to look at it and, if necessary, melt more solder to flow inside and assure a better connection. Be sure not to attempt to bend or pinch the hollow contact pin in any way but simply rely on heat and solder flow to repair this particular defect. Of course, replacing the tube with a new one is an easy solution, but where a replacement tube is not available or is quite expensive, as in the case of TV picture tubes, the repair effort is well worthwhile.

5.4 HOW TO USE THE SYMPTOM-FUNCTION TECHNIQUE IN TUBE CIRCUITS

CAUTION—TUBES ARE HOT, VOLTAGES ARE HIGH. WATCH WHAT YOU TOUCH—KEEP ONE HAND IN THE POCKET.

The above warning applies to all tube troubleshooting and cannot be stressed too much. Even minor burns reduce your effectiveness and, in that sense, cost money. Electrical shocks can be really dangerous.

The basic approach described in chapters 3 and 4 for using the symptom-function technique in transistors and integrated circuits is

119

the same for vacuum tubes. In many respects, however, the fact that vacuum tubes are easily removed and replaced makes this technique even more useful. Suppose, in a color TV set, only the sound is intermittent or distorted. We can quickly tap the vacuum tubes responsible for the audio portion, one by one. We can jiggle the tubes—use a glove or a rag to avoid getting your fingers burnt—and chances are that we will find the one causing the defect. Because the substitution technique is so easily used, we frequently combine it with the symptom-function technique and replace tubes in the suspected portions of the equipment. In such equipment as radios, stereo systems, TV sets, and other consumer items, most experienced troubleshooters find that 80 to 90% of all their troubles can be solved in this manner. In the remaining cases we must use the signal-tracing technique or the voltage-resistance technique.

5.5 HOW TO USE SIGNAL TRACING IN TUBE CIRCUITS

CAREFUL—CAREFUL!
ELECTRIC SHOCKS AND NASTY BURNS CAN COME FROM CARELESS TUBE TROUBLESHOOTING

When the signal-tracing technique is applied to vacuum tube circuits, a few important characteristics must be kept in mind. In vacuum tubes, in general, the input impedance at the control grid is very high, usually in the order of one megohm or more. This means that when the output of the signal generator is applied to the control grid, this grid will inevitably be loaded down and the gain of that stage, at least, will be greatly reduced. The output impedance of vacuum tubes is generally in the order of 5,000 to 50,000 ohms. Fortunately, VTVMs and oscilloscopes have high input impedance and can therefore be used without special problems.

Remember that the plate voltage of vacuum tubes is at least in the order of 100 volts or more. The signals that we usually want to measure are much smaller than such a high voltage and we must therefore isolate the signal to be measured from the DC power voltage. Capacitive coupling is an obvious solution but we must be sure that the capacity is large enough and does not represent a significant series impedance that could change the observed signal. Most oscilloscopes have both an AC and a DC input. When working with vacuum tubes, the input selector switch on the oscilloscope should invariably be set to AC. This will switch in an internal isolating capacitor. With the exception of transmitting tubes and other high-power circuits, the currents supplied to vacuum tubes are relatively small, but if you touch your finger to anything over 100 volts you are very likely to get a strong shock. It can be really dangerous if your body is grounded. Many experienced troubleshooters automatically keep one hand in their pocket when probing around in a live vacuum tube chassis.

In signal tracing vacuum tubes, some techniques are possible that cannot be used with transistors or ICs. In audio circuits, for example, it is easy to use a clip lead and ground grids and cathodes of successive stages until you hear the click in the loudspeaker and thereby assure yourself that a particular stage still works. In troubleshooting RF circuits, it is possible to couple RF signals from a signal generator directly into the tube by placing a loop of a few turns of wire around the glass tube envelope and connecting this loop to the output of the signal generator. This method is frequently used to inject RF and IF signals into FM and TV receivers where sufficient amplification is provided so that the relatively weak RF signal from the generator will appear as a test pattern or as a tone at the picture tube or the loudspeaker, respectively. Where relatively large power amplification is used, such as in RF transmitters, diathermy machines, or TV horizontal flyback circuits, it is possible to use a loop connected to the oscilloscope probe to determine whether signals are present at the tube itself. This approach is particularly useful when high voltage makes it inconvenient to use the probe directly at the vacuum tube terminals.

Just like transistors, tubes require the correct bias. In most circuits the grid bias is provided by means of the cathode resistor and capacitor, but in many instances, particularly in IF and RF amplifier circuits, a separate AGC (automatic gain control) bias is applied to the grid. In signal tracing tube circuits it may be necessary to substitute the appropriate grid bias in place of the AGC bias by means of a set of batteries and a potentiometer. Typical bias values range in the order of -3 volts. Remember to look at the circuit diagram before substituting grid bias and to check the various R-C decoupling networks that isolate the individual tube control grid from the bias bus. In low-level RF amplifiers all of the B+ voltages and very often the filaments as well are decoupled to prevent RF feedback. Depending on the particular circuit, the decoupling networks may consist of coils and capacitors as well as the standard R-C network.

121

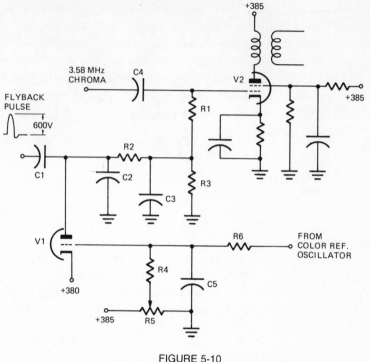

FIGURE 5-10
SIMPLIFIED COLOR-KILLER CIRCUIT

To illustrate signal tracing in tube circuits, refer to Figure 5-10, which shows the simplified color-killer circuit found in many earlier color TV receivers. The tetrode V2 is the amplifier for the 3.58 MHz chroma subcarrier and receives its signal through C4 at the control grid. A conventional cathode bias circuit is used and the output of the tube goes to a transformer. When color transmissions are received, V2 acts as a normal, class A1 band pass amplifier. When black and white transmissions are received, V1 generates a negative bias, applied to the control grid of the V2 through R1, which cuts this tube off, preventing noise signals from being demodulated and appearing as color. If the color-killer circuit does not work on black and white transmission, the black and white picture will be interrupted by red, green, and blue dots flashing randomly all over the screen. If V1 operates incorrectly, V2 could be cut off even during color transmissions and these would then appear in black and white. As indicated by the semicircles around V1 and V2, each of these tubes is part of

122

another tube envelope and simple tube substitution is therefore not very practical.

V1 receives its plate voltage through C1 in the form of a +600 volt portion of the horizontal flyback pulse. Depending on the voltage at the grid of V1, V1 will conduct or be cut off during the flyback pulse period. The plate current, if the grid is not too negative, flows through R2 and R3 and sets up a negative bias at the junction of R1 and R3, adding to the cathode bias of V2. If the flyback pulse is absent or too weak, this bias will not appear and V2 will never be cut off. The control grid of V1 receives a 3.58 MHz signal from the color reference oscillator and this signal will only appear if the color sync burst has been demodulated from the horizontal synchronizing pulse. The operating point of the grid of V1 is determined by the potentiometer R5. Since the cathode of V1 is returned to +380 volts, it is obviously possible to set the control grid at +5 volts, causing grid current to flow. It is also possible to cut off V1 completely by setting the potentiometer to a sufficiently negative point.

In signal tracing the simplified color-killer circuit shown in Figure 5-10, it is not necessary to inject a test signal and the VTVM will not be of much help. If we can use a wideband oscilloscope, we can put the probe at either side of R6 to make sure that the 3.58 MHz signal from the color reference oscillator is present during a color transmission. We can also check the presence of the chroma signal at the grid of V2. As a third step we should check, on the plate of V1, whether the flyback pulse is present at that point. When we short out R1, there should be no color, even though color transmission is received. If we vary the potentiometer so that the DC grid voltage is close to the cathode potential, we should also observe that the voltage across R3 becomes increasingly negative. If this does not happen, V1 can be suspected. We can next check across C2 and C3 to make sure that the flyback pulse, somewhat reduced at C3, is still present. With the DC voltages at grid and cathode of V1 correct and with a 600-volt flyback pulse applied to the plate of V1, and with no current passing through R2 and R3, it becomes highly likely that V1 is defective.

In our experience with this type of circuit, we have often found that C2 has been shorted, causing the same symptoms as if V1 were defective. It has also happened that C1 was open or shorted, causing V1 to short or else V1 not to receive any plate voltage. Signal tracing the circuit with the oscilloscope inevitably will find the defective part.

5.6 HOW TO USE THE VOLTAGE-RESISTANCE TECHNIQUE IN TUBE CIRCUITS

REMEMBER: TUBES ARE HOT, VOLTAGES ARE HIGH. BE CAREFUL WITH YOUR HANDS.

Because vacuum tubes are voltage devices rather than current devices, such as the transistors and ICs are, the voltage-measuring technique got its start in the early days of electronics when vacuum tubes were the only available active circuit elements. It is customary for equipment manufacturers to print the voltage values for cathode, grid, screen, suppressor, and plate for all tubes, usually directly on the circuit diagram. This is a great help to us because, with only a VTVM, we can quickly measure voltages on the operating equipment and very often find the defective component right away. In a typical circuit, for example, we will measure the voltages at the tube sockets and we may find that there is no screen voltage. Checking at the other end of the screen resistor will tell us whether that resistor itself may be open. If a capacitor is used to bypass the screen to ground, it may be that this capacitor is shorted. We just turn off the power and check, with the ohmmeter, which of the two components is open or shorted respectively. With the power removed from the equipment it is possible to check resistance values to B+ and to ground at all of the tube pins and thereby quickly determine if any of the passive components are open, shorted, or have the wrong values. Figure 5-11 shows a typical voltage and resistance chart, as supplied by the manufacturer. We can see right away that it would not be efficient to simply measure all voltages and resistances since this would take quite a bit of time. Only after the symptom-function technique or the signal-tracing technique has isolated the fault to a particular portion of the equipment is it worthwhile to refer to the voltage-resistance chart and make the required measurements.

When we encounter a defect that seems particularly hard to troubleshoot, possibly because the defect is intermittent or occurs only after the equipment gets hot, it is a good idea to write down the voltage and resistance measurements taken at different times and under different conditions of temperature so that we can compare them later. This is particularly important in cases where a replacement tube seems to work well for a certain time but then seems to become defective again.

RESISTANCE CHART

PIN NO.

Tube	1	2	3	4	5	6	7	8	9	10	11	12
V1 6AU8	47K	12.9K*	3K**	NS F	NS F	800 to 2.8K	1M	18K***	12.8K*			
V2 6AU8	47K	12.9K*	3K**	NS F	NS F	800 to 2.8K	100	18K***	12.8K*			
V3 12BY7	1.8K to 2.5K	47K	1.8K to 2.5K	NS F	NS F	NS F	4.17K*	11K*	1.8K to 2.3K			
V4 12BY7	1.8K to 2.5K	47K	1.8K to 2.5K	NS F	NS F	NS F	4.17K*	11K*	1.8K to 2.3K			
V5 6BL8	16K**	150K	136K**	NS F	NS F	4.5K	0	2.5K*	2.2M			
V6 12AZ7	48K*	100 to 10K	6.8K	NS F	NS F	48K*	10.5K to 45K	6.8K*	NS F			
V7 12AZ7	27K*	680K* to 11M	4.7K	NS F	NS F	10K*	20.5K*	4.7K	NS F			
V8 EZ81	100	—	0	NS F	NS F	—	100	—	—			
V9 IV2	—	—	—	750	750	—	—	—	6M*			
V10 WX5013	5.95*	6M*	5.78M to 6.28M	3.3M to 5.3M		4.17K*	4.17K*	0* to 2M	48K*	48K*	—	5.8K*

<u>REFERENCES FOR RESISTANCE CHART</u>

IMPORTANT: All resistance measurements in table are made with pin 3 of EZ81 power rectifier V8 (B+) and can of C26 (B-), located in Fig. 7, <u>shorted to ground with temporary jumper leads. Remove these B+ and B- shorts to check resistance at pin 3 of V8</u> (should be <u>greater than 70K ohms</u>) and before connecting unit to power line.

*Small delay until final reading can be obtained due to capacitor charging.
**Large delay until final reading can be obtained due to capacitor charging.
***Depends on setting of DC balance adjust R13.
N.S. — Not significant
F   — Filament pin
All resistance values may normally vary by ±15%.

Courtesy Eico

FIGURE 5-11a
RESISTANCE CHART, EICO 435 OSCILLOSCOPE

VOLTAGE CHART

PIN NO.

TYPE	1	2	3	4	5	6	7	8	9	10	11	12
V1 6AU8	93V*	90V*	280V	3.15V AC	3.15V AC	2.5V	0	112V	90V*	—	—	—
V2 6AU8	93V*	90V*	280V	3.15V AC	3.15V AC	2.5V	0	112V	90V*	—	—	—
V3 12BY7	95V*	93V*	95V*	3.15V AC	3.15V AC	3.15V AC	270V*	240V	95V*	—	—	—
V4 12BY7	95V*	93V*	95V* AC	3.15V AC	3.15V AC	3.15V AC	270V*	240V	95V*	—	—	—
V5 6BL8	200V	0***	47V	3.15V AC	3.15V AC	5V	0	20V	12V***	—	—	—
V6 12AZ7	200V*	0	1.7V	3.15V AC	3.15V AC	200V*	-15V to +10V	1.7V	3.15V AC	—	—	—
V7 12AZ7	130V	50V	65V	3.15V AC	3.15V AC	200V	64V	65V	3.15V AC	—	—	—
V8 EZ81	360V AC	—	430V	—	—	—	360V AC	—	—	—	—	—
V9 IV2	—	—	—	1070V AC ****	1070V AC ****	—	—	—	-1600V	—	—	—
V10 WX5013	-1500V	-1500V	-1500V to -1300V	-760V to -1200V	—	270V*	270V*	0 to 360V	200V*	200V*	—	-1500V

REFERENCES FOR VOLTAGE CHART

*Varies with VERT. POS./HORIZ. control settings; center values given.
**Goes negative with vertical input signal applied at SYNC "+" or "-" positions of HORIZ. SELECTOR.
***Difficult to measure at this point. Make measurement at junction of R51 and R52.
****Measure with tube V9 (IV2) out of socket, each pin to ground. Do not attempt to measure filament
     voltage between pins 4 and 5 of V9.
UNLESS OTHERWISE INDICATED, ALL VOLTAGES ARE DC, POSITIVE AND MEASURED TO CHASSIS.
Line Voltage: 117V, 60 cps
All measurements made with VTVM of approximately 11 megs input impedance.
All voltages may normally vary by ±15%.

Courtesy Eico

FIGURE 5-11b
VOLTAGE CHART, EICO 435 OSCILLOSCOPE

Finding Intermittent
Defects

6.1 WHAT CAN CAUSE AN INTERMITTENT DEFECT

One of the most frustrating experiences is to have a toothache disappear just as we reach the dentist's office. We immediately feel like turning around, but we know that the toothache will come back. The dentist is now confronted with a problem very similar to the intermittent defects in electronic equipment. Usually the dentist looks at the suspected tooth and pokes around until he finds the cavity. If it is not obvious, he will take an x-ray picture. In a few rare cases the dentist has to resort to other methods, such as testing with hot and cold water, with the air-jet, or with an electric stimulator. Intermittent defects in electronic equipment are sometimes easy to find and sometimes they seem to present an impossible problem. The vast majority of intermittent defects, however, just like a toothache, will yield to a careful and methodical troubleshooting procedure.

It often helps to remember that intermittent defects are basically no different from the other troubles discussed in the previous five chapters. What makes intermittents so difficult to locate is their very nature—they are not always present.

The first question we ask ourselves about intermittent defects is: What makes them intermittent? Invariably the answer is that an electrical connection, somewhere in the circuit, opens or shorts intermittently. Two basic defects cause the intermittent characteristics. One is a mechanical change, such as vibration, shock, position of the equipment (the force of gravity), air-flow, etc. The second major cause of intermittents is a change in temperature. An unequal expansion or contraction of two materials can cause bending or twisting, which then results in a shorted or open connection. A third element, less frequent in electronic equipment, is moisture. Excessive moisture leads to poor insulation and this can cause intermittent voltage breakdowns in some circuits. The following partial list of typical intermittent types of defects will give you a better clue as to what to look for when you encounter this type of trouble.

(a) *Poor connector contact*—Always suspected when any sort of connectors are involved in the equipment. Figure 6-1 illustrates the case where the receptacle, or jack, does not make complete contact with the plug portion. Any mechanical vibration or other motion can cause the plug to be loose in the receptacle and contact will be lost.

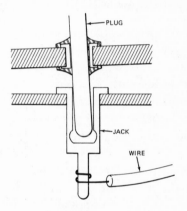

FIGURE 6-1
LOOSE CONNECTOR CONTACT

(b) *Broken wires*—Quite frequently the cause of intermittent defects. Remember that most connectors are used to connect wires with other wires and, in addition to the contacts from the connectors themselves, the wires leading to these contacts can be defective. Figure 6-2 shows a particularly nasty, but typical, case of intermit-

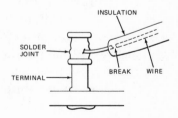

FIGURE 6-2
"INVISIBLE" BROKEN WIRE

128

tent. The break in the wire is concealed by the insulation. Actually the defect was started when the wire stripper pinched the wire at the point where the insulation was cut off. Vibration between the solder joint and the wire itself caused the break. The insulation has since covered the pinched spot, keeping the wire in place and concealing the intermittent point of the connection.

(c) *Loose metal*—Such as pieces of solder, cut-off pieces of wire, etc., can move around the chassis of the electronic equipment and lodge themselves between two connections in such a way as to cause an intermittent short circuit.

(d) *Bad solder joints*—Can occur in any type of equipment and are a frequent source of intermittent contacts. Even though three or four wires may seem solidly soldered together, corrosion, chemical contamination, or other effects may have caused one of the wires to make only an intermittent connection within the solder joint.

(e) *Loose ground connection*—Often due to corrosion and loosening of the pressure contact. Many ground connections consist of a solder lug screwed against the chassis. When the screw loosens and corrosion creeps in between the solder lug and the chassis itself, an intermittent connection can occur. Intermittent RF grounds are particularly difficult to find since they cannot be detected by troubleshooting with the equipment turned off.

(f) *Broken elements in tubes or transistors*—Also includes internal defects in integrated circuits and other active assemblies. Once this type of defect is traced to the functional circuit, a simple replacement of the tube, transistor, or IC will cure the problem.

(g) *Defective insulation between conductors*—Can range from such obvious faults as a cracked insulator in an automobile radio antenna to such subtle trouble as internal arcing in a transformer due to deteriorated insulation between windings. Some types of insulating materials dry up, become brittle, or otherwise deteriorate with time. Other types of insulating materials may melt due to excessive heat (such as wax) or else can crack, tear, or otherwise be damaged. The intermittent defect is due to a short circuit that the insulation sometimes cannot prevent.

(h) *Printed circuit board intermittents*—Can range from a cracked PC board, and the resultant intermittents in the conductors, to short circuits between conductors, or a defective, plated-through hole. Figure 6-3 shows how a short circuit between conductors can occur by buildup of dirt, metal filings, or tiny solder pieces bridging two conductors on a PC board. The problem with intermittent plated-through holes is usually more complex and Figure 6-4 illustrates, in cross section, how a break can occur between the two sides of a plated-through hole.

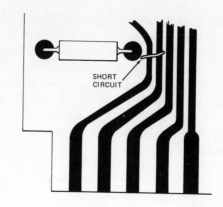

FIGURE 6-3
SHORT CIRCUIT ON PC BOARD

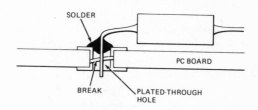

FIGURE 6-4
BREAK IN PLATED-THROUGH HOLE

(i) *A cracked or broken tuning slug in a variable inductor*—May be rare but is difficult to find. When this happens in a TV tuner, the defect may cause grey hairs in even the most experienced service technician.

(j) *Loose lamination or a cracked core in a transformer*—Such as the flyback and high-voltage transformer in a TV set, can also account for strange defects. Once this type of intermittent is traced to the defective transformer, it has to be replaced since repairs are not usually possible.

The short list of typical intermittent defects above indicates that almost anything can account for an intermittent. Anything that can break, short, or change with vibration, shock, temperature change, etc., can cause an intermittent. We shall see later in this chapter how we find each of the types of intermittent circuits relatively simply and quickly.

130

6.2 ISOLATING THE INTERMITTENT DEFECT TO ONE CIRCUIT

In the old vacuum-tube radio and TV receivers, the first approach to locating any kind of intermittent defect was to make sure that all tubes were firmly seated in their sockets and then to tap each one with a screwdriver handle. If this produced the intermittent defect, the tube was exchanged and the trouble was cured. In present-day electronic equipment, this approach is no longer practical and a more sophisticated method of isolating the intermittent defect to a particular circuit has to be used.

The four basic methods of troubleshooting were described in some detail in chapter 1, and chapters 3, 4, and 5 showed you how these methods can be applied to troubleshooting transistor circuits, integrated circuits, and electron tube circuits. You have seen how the symptom-function technique lets you do a lot of troubleshooting simply by analyzing the symptoms and understanding what portions of the equipment can cause such a symptom. In isolating intermittent defects, this method is almost always the first step. If, in a TV set, the high voltage seems to arc or go on and off intermittently, we would be wasting our time to look for trouble in the TV tuner or the IF section. Obviously, the high-voltage section is the first suspect. Similarly, when the frequency response of the hi-fi system seems to change intermittently and the treble and base varies unpredictably, we will look for the intermittent defect in the tone-control section. You can save a lot of troubleshooting time in intermittent defects if you can reason out, before ever opening up the equipment, in which portion the defect is most likely to be. Consult a block diagram or a circuit diagram, and observe the symptoms carefully first.

The substitution technique is usually only useful after we have already narrowed down the intermittent defect to one or a few suspected parts. The signal-tracing and the voltage-resistance techniques usually do not help a great deal in isolating intermittent defects to one circuit. There are exceptions to this, but, as a rule, it is the symptom-function technique that really helps you to isolate the intermittent defect.

The last part of this chapter will illustrate a method that can be used as a last resort to change intermittent defects to permanent defects. Permanent defects can, of course, be located by the regular troubleshooting methods described elsewhere in this book.

6.3 LOOKING FOR INTERMITTENTS

Once we know that an intermittent defect exists, we will use the symptom-function technique to get some idea of, at least, in which particular portion of the equipment the intermittent defect is located.

Before taking the equipment apart, however, it's a good idea to jar, bang, or shake the entire chassis while looking for the symptom. If the jarring, banging, or shaking causes the defect to either appear or disappear, then we know that some kind of mechanical intermittent exists. If we can observe the symptom closely during the jarring, we may even localize the trouble further. In a TV set, for example, if the raster remains steady but the picture seems to break up, the most likely defect is in the video or IF portion and not in the horizontal sync or high-voltage circuits. If the picture is good throughout the jarring, banging, and shaking, and only the sound is affected, then we know that the intermittent is in the audio section.

Now we come to the crucial point—we have to take the equipment apart and look for the defect.

Because of the possibility that a loose piece of metal has caused the intermittent to begin with, experienced service technicians make sure that when they disassemble the equipment, take the chassis out of the cabinet or remove backplates, etc., they will know if any loose piece of metal falls out. With the AC power removed, the open chassis or assembly should be shaken out, preferably over a clean surface, so that anything that falls out can be seen. In most electronic equipment dust and dirt will have accumulated inside but metal parts can usually be spotted easily. Well-equiped service shops usually have either an air-compressor or a spray can capable of furnishing a strong jet of clean, dry air, such as "Dry-Blow" manufactured by H.M.M. Industries, and available with an 18-inch-long extender tube. This permits forcing air under wires, components, etc., where a loose piece of metal may have lodged. After blowing out the assembly, turn it upside down and shake it lightly to allow any trapped metal to fall out. Even if you haven't seen any loose metal so far, it may be a good idea to connect AC power and turn the equipment on to see whether the intermittent defect will exists. Slightly jarring the entire chassis or assembly should verify that the defect is still there.

With the AC power off, a visual inspection of the suspected portion of the equipment, according to the symptom-function technique application mentioned above, is the next step. A good light, a pointed tool, a small mirror, and possibly even an enlarging glass, are the only requirements for a detailed visual inspection. Any discoloration, flow of wax, or some other insulation material, indicates a possibly defective part. We have seen discolored carbon resistors that measure correctly but, when tapped with a screwdriver, break apart. They have obviously been overheated and are making only an intermittent contact.

If the visual inspection does not show up the intermittent defect, the next step is to turn the power on and try to produce the intermittent defect symptom by bending, pulling, tapping and jiggling wires, components, and terminals in the suspected circuit portion. A good way to do this is to use an insulated tool, like some of the alignment tools now on the market, or else to use a forked hardwood or plastic stick, as illustrated in Figure 6-5. The idea of this procedure is to reach a particular wire, component, or terminal that, when you start putting stress on it, will cause the total failure that was previously only intermittent. Look for terminals that are close together and see whether bending or tapping the chassis does not tip them sufficiently so that they short. Look for PC boards that, when bent one way or the other, may cause the intermittent defect. Look for connectors, sockets, and other plug-in devices to make sure that jiggling or bending them or their mounting surface does not cause an intermittent. A particularly frequent source of intermittents are ground lugs, usually the type that is screwed onto the chassis and where several wires are soldered to it. The solder joint may be bad but most likely the screw and nut have become loose. Riveted solder lugs are even more likely to be the source of intermittent defects. Try all of them in the chassis by pushing and pulling on them while the power is on.

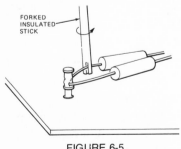

FORKED
INSULATED
STICK

FIGURE 6-5
TESTING FOR INTERMITTENT
SOLDERED CONNECTION

133

One of the most powerful ways of finding intermittent components, hairline cracks in PC boards, and similar types of intermittent defects is to use the instant freezing type of spray can. These cans, such as the "Super Frost 8" manufactured by Chemtronics, or the "Propellon Instant Freeze," manufactured by H.M.H. Industries, are available with extender tubes that let you pinpoint the cold spot to the suspected portion of the circuit. Metals contract in the cold and this usually makes the intermittent defect more apparent. Be careful in using these freezing sprays, however, because many semiconductors do not operate at very low temperatures and may be damaged if the freezing spray is applied directly.

The opposite approach is to heat up the suspected area, usually by means of a heat gun or by holding a strong soldering iron close to the suspected component. Remember again that semiconductors, particularly those using germanium, can be damaged by excessive heat. The use of the heat gun or soldering iron for testing high-temperature intermittents should be limited to those cases where the intermittent defect only appears after a considerable warm-up period. A detailed discussion of finding some of the most frequent types of intermittents is found in the following paragraphs.

6.4 HOW TO FIND A BAD SOLDER JOINT

The electronic equipment going into the Minuteman missile system had to be ultrareliable. An attempt was made to train individuals to inspect solder joints to make sure that they were "good." Inspectors were equipped with magnifying glasses and with strong lights. Solder joints that were inspected and found satisfactory were painted off with a purple marker. In subsequent vibration, shock, temperature, and other environmental tests, the failures were analyzed. Many failures were due to defective solder joints even though they had been inspected and checked by carefully trained individuals. The moral of the story is that it is very difficult to spot a defective solder joint.

Fortunately, however, a method exists to cure intermittent defects that may be caused by bad solder joints. This method is relatively simple. It just means that every solder joint in the suspected circuit is resoldered. When you find, in troubleshooting intermittent defects in a digital counter, that the most significant bit seems to be the one affected, you will first look at the circuit and components for obvious breaks and similar defects that could cause an intermittent. If you cannot find them right away, it is a very good practice to simply resolder all joints, one by one, that may affect the most significant bit.

A few words about resoldering. The size of the solder joint determines the size of the iron to use. If it's an integrated circuit mounted on a PC board, it would be wrong to use a 100-watt iron. A 15- or 25-watt pencil-type iron will be the right size. If we are dealing with a CB transmitter, where we suspect the wiring on one of the transmitter tube sockets, a 100- or 200-watt iron will be the right thing to use. Make sure that the solder flows smoothly. Fresh solder, preferably of the rosin-core type, should be used where the old solder joint is not completely filled with solder. Be sure to let the solder joint cool by itself and do not try to shock it into hardening by applying a screwdriver or pliers to it before it has hardened by itself. It is usually more efficient to resolder all of the connections within the suspected circuit than to turn the equipment on after each individual resoldering.

6.5 HOW TO FIND A BAD CONNECTION ON A PC BOARD

Intermittents on PC boards, as briefly mentioned in paragraph 6.1 above, are usually due to either a crack that causes an open circuit, a short between conductors, or a defective plated-through hole. A short between conductors is usually the easiest to find. You may need an enlarging lens to see the actual short circuit, because it may be a hair-thin wire, a piece of solder, or even the so-called "whiskers," a form of crystal structure bridging between adjacent conductors. Without an enlarging glass, one simple and relatively quick solution to the shorting problem is to run a pointed tool, such as a scriber, awl, or ice pick, gently between the adjacent conductors where they are spaced relatively close together. Another approach is to wash the suspected area of the PC board with a good cleaning fluid, making sure that it dries cleanly afterwards. Drying with an air-jet also helps.

A cracked or broken conductor is much harder to find. Often, the break is very thin and may only be apparent when the circuit board is under some kind of tension or stress. The best method to find this type of intermittent defect is to jumper all suspected conductors. Connect a piece of insulated wire from one point on the conductor to the other point, such as the land for a resistor, capacitor, transistor, etc. It is usually more efficient to jumper several suspected conductors at one time and then see if the intermittent persists, than to turn the equipment on after putting each jumper in place individually. If the intermittent defect is now cured, a repair has already been made by the installation of the jumpers. If it is not practical, for some reason, to use wire jumpers, it is possible to coat the entire suspected conductor with solder, by very carefully using a small soldering iron, and this may bridge the intermittent open circuit. When this method is used, be sure to check that in installing the solder coating you have not also installed a short circuit, intermittent or permanent.

To find a defective plated-through hole is very difficult but, as in the approach to a bad solder joint, the simplest solution is to resolder all plated-through holes in the suspected circuit. Referring to the sketch in Figure 6-4, this is accomplished by resoldering both sides of the plated-through holes and making sure that the wire, in this case, the lead from a component, is properly soldered on both sides. The wire will then furnish the connection, regardless of the internal situation in the through-plated hole.

A much less frequent source of PC board intermittents are the edge connectors. Edge connectors on the printed circuit board itself can be inspected to insure that they are in good condition and that they neither short nor are coated with dirt or grease. In the next paragraph, we will show you how to find a bad connection due to the connector itself. PC board troubles are often due to the connector rather than the PC board itself.

6.6 HOW TO FIND A BAD CONNECTION ON A CONNECTOR

One of the most interesting results of the Minuteman reliability program mentioned above was the finding that most of the defects, particularly intermittent defects, were caused by connectors. This was especially unfortunate because the modular design requirements of the Minuteman equipment resulted in the use of thousands and thousands of connectors. High-reliability connectors were developed in which the contacts between the plug and jack portion were outstanding examples of metallurgical science. Rhodium, nickel, gold, etc., were all used in various plating arrangements to assure a noncorrosive, properly mating, large-surface contact. In ordinary equipment, particularly those made to commercial standards, the connectors are not nearly as reliable as those of the Minuteman program and connectors still rank as one of the major causes for intermittent defects. Whenever you suspect an intermittent connection, the first thing to do is to separate the connectors, plug and jack portions, and look for broken parts, dirt, dust, foreign material, etc. If a strong jet of air is available, it is a good idea to "blow out" both portions of the connector, and to try again to see whether the intermittent defect is still there. Check the wired connections to both the jack and the plug side. Frequently, the intermittent defect is at the point where a cable or a lead is soldered to the terminal of the connector itself.

The ultimate way to find the intermittent in a connector itself is to bridge the connector, usually by means of a clip lead or else by a wire soldered across. Rather than simply trying to jumper every one of the pins and sockets of a multipin connector, try to reason out by the nature of the intermittent defect, in an application of the symptom-function method, which wire, which pin and socket, which jack and plug portion, is most likely to be at fault. In some connectors it is possible to replace individual pins and that would then constitute the repair. In many other connectors, however, the entire connector has to be replaced if it is defective.

6.7 A "LAST-RESORT" METHOD

Sometimes it happens, no matter what we try, that the intermittent defect cannot be found. We may go over the suspected intermittent connections, components, etc., point by point, we may seem to use a fine-tooth comb to catch any kind of short or open circuit, and yet —every time that we think we have repaired the defect—it shows up again just as soon as the cabinet is reassembled. The last-resort method described here is adapted from the reliability testing approach used by the military services to determine just how good electronic equipment is.

Figure 6-6 illustrates the relatively simple case of an FM stereo tuner that had an intermittent defect we were unable to find, even after trying every one of the steps recommended in the preceding paragraphs. We wrapped the tuner itself in a piece of old blanket and strapped it to a half-inch plywood board. Next, we put a triangular piece of molding under the board to set up a sort of seesaw arrangement. We turned the unit on, let it warm up for ten minutes, and

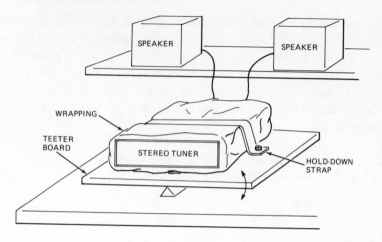

FIGURE 6-6
HEAT, SHOCK, AND VIBRATION
TEST FOR A STEREO TUNER

then pushed one edge of the mounting board up and down while observing the intermittent defect. The combination of the heat build-up in the unventilated stereo tuner and the shock and vibration due to the teeter board action eventually caused a failure in one of the stereo channels. We removed the stereo tuner from the mounting board, unwrapped it, let it cool off, and turned it on again. The intermittent defect had become permanent and was then signal-traced without much difficulty to a defective coupling capacitor.

Another application of the last-resort method concerned a TV set that we covered with a blanket to increase the heat build-up. Vibration was applied by tapping a corner of the chassis with a rawhide hammer at approximately one-second intervals. In this case, the horizontal output tube failed first. It was replaced, the process was repeated, and eventually the high-voltage capacitor became shorted. This apparently had caused the intermittent defect because when the capacitor was replaced, the intermittent defect disappeared.

Just how you apply the last-resort method to a particular piece of equipment you are trying to repair depends largely on the equipment and on your ingenuity. If the equipment is bulky and heavy, you cannot easily use the teeter board method illustrated in Figure 6-6. If it contains forced-air cooling, with air pressure switches that turn off the equipment when the cooling is blocked, you must find some other way to increase the temperature. In any event, the key to the last-resort method is to stress the electronic equipment by applying both heat and mechanical vibration at the same time. A combination of these two effects usually causes the marginal or intermittent components to fail first. Once the intermittent defect has become permanent, you can find it much easier.

7

How to Deal with Interference Defects

7.1 WHAT CAN CAUSE INTERFERENCE DEFECTS

When we watch the all-important Sunday football game and a female voice shoots a series of important questions at us just as the announcer is explaining a particular play, we know that we suffer from interference. In this case it is audio interference and it is an external interference, unless the female voice comes from within the TV set. In dealing with interference defects in electronic equipment we also recognize the difference between external and internal interference and we also recognize the importance of determining just what kind of interference it is before we can try to fix it.

In many kinds of electronic equipment it is difficult to recognize interference as such. If the voice in a stereo set sounds garbled, is it interference or defective synchronization? If the mini-computer consistently prints out wrong numbers, is it due to some defect or due to noise causing false triggering in some counting circuit? If we look carefully at the different causes of interference and how they affect the normal operation of equipment, we will be able to find out quickly whether the defect is indeed due to interference and whether the interference is from the outside or within the equipment.

One characteristic common to every type of interference is that it is always in addition to or mixed with the normal signal flow. For example, the most frequent type of outside interference, static noise, is received together with the desired RF signal. Other types of interference, such as a radio amateur transmitter interfering with TV reception, are simply additional electronic signals added to the existing one. AC hum, whether due to the power supply of the equipment or due to some other, external AC source, is another type of external

140

interference. Sometimes it is difficult to define what is "external" because, while an interference signal may be external to the circuit with which it interferes, it may originate in, or be amplified by, some other circuit internal to the overall equipment. Power supply hum is typical of this type of interference.

Three fundamentally different types of interference can occur. In the first case, the interfering signal is heterodyned with the desired signal. Figure 7-1 shows the basic hetrodyning principle. Two fre-

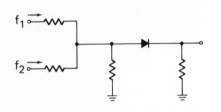

FUNDAMENTALS

$f_1, f_2, (f_1 + f_2), (f_1 - f_2)$

HARMONICS:

nf_1, mf_2

$(nf_1 + mf_2), (nf_1 - mf_2)$

FIGURE 7-1
PRINCIPLES OF HETERODYNING

quencies, f_1 and f_2, are mixed in a nonlinear impedance, in this case a simple diode. A nonlinear impedance can be anything, whether it is a diode, a portion of a transistor, a vacuum tube, or even a poor solder joint. As long as the impedance is nonlinear with input amplitude, the output of the circuit will consist of four fundamental frequencies and, depending on the circuit, a large number of harmonics. The fundamentals are the two original frequencies, their sum and their difference. As illustrated in Figure 7-1, the number of harmonics that may result from this heterodyning process can be quite large. A great deal depends on the frequency response of the output of this nonlinear impedance circuit. If the frequency response is limited to the lower portions, only f_1 minus f_2 may be passed. This is the case in a simple AM radio where f_1 may be the local oscillator frequency and f_2 the desired incoming signal. The output, $f_1 - f_2$, is usually 455 KHz, the IF frequency used in AM receivers. Since f_1, f_2, and, of course, the sum of the two, are much higher than 455 KHz, they are not amplified and are essentially attenuated at the output of the mixer. In this type of AM receiver, both the incoming signal and the local oscillator are relatively weak signals and their harmonics, above and below, are also weak.

141

In a typical example of heterodyne, or beat, interference in TV reception, a TV set tuned to channel 4 (66 to 72 MHz) also received interference from a nearby radio amateur transmitter. We can often determine where this interference originates because we can hear the radio amateur's voice or else pick up his Morse code signals and identify him in that way. We notice this type of interference as streaks going through the picture whenever he goes on the air. The question is, where does the interference enter? If the amateur operates at 3.5 MHz, it is possible to pick up this frequency in the video channel where it is amplified along with the 60 Hz to 4 MHz video signal. If the amateur transmitter is reasonably close by and radiates with sufficient output power, the 12th harmonic of 3.5 MHz would be 42 MHz. The IF frequency band of standard TV receivers extends from 41.25 to 45.75 MHz and could therefore pick up the amateur and amplify the signal to cause the apparent beat. The third possibility is that the 20th harmonic of 3.5 MHz interferes directly with the RF and is picked up by the antenna. This would place the interfering frequency at 70 MHz, sufficiently in the band of channel 4 to cause severe interference. In a later paragraph we show you how to find out where the interference comes in and what to do about it.

The second type of interference is illustrated in Figure 7-2, which shows a sinewave signal being superimposed by the interfering signal. In this case, the two frequencies are simply added together. This type of interference is typical of hum pickup, which may ride in over the

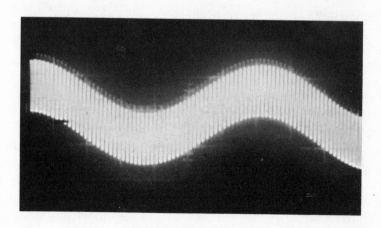

FIGURE 7-2
INTERFERENCE ON SINEWAVE

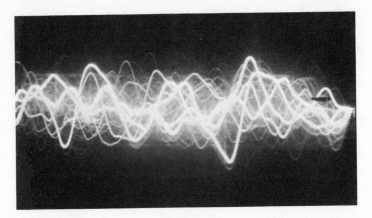

FIGURE 7-3
NOISE

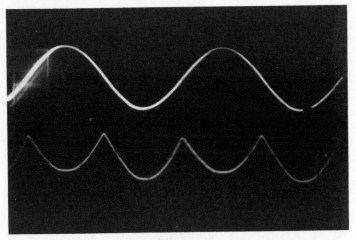

FIGURE 7-4
60 Hz SINEWAVE AND 120 Hz RECTIFIED SIGNAL

audio signal. Another frequently found interference that is usually added to the desired signal is so-called "white noise." The appearance of noise by itself is shown in the oscilloscope photograph in Figure 7-3, while the appearance of 60 Hz and 120 Hz hum is shown in Figure 7-4. If you thought that hum is always 60 Hz, you are wrong in most cases. The 60 Hz signal gets changed to 120 Hz by the action of a full-wave rectifier or a bridge rectifier, as in Figure 7-4, the most common types employed in power supplies for electronic equipment.

The third type of basic interference signal is due to internal oscillation. This may be a linear or sinewave type of oscillation or it may be a relaxation type of oscillation. Most likely the oscillation is due to feedback, usually in a tuned amplifier, and this presents special problems that are discussed in more detail in paragraph 7.3 below.

7.2 SEPARATING INTERNAL FROM EXTERNAL INTERFERENCE

Experienced troubleshooters first ask the one critical question: "Is it always there?" If the answer is no, the interference is not always there, then it is most likely due to some external cause. If the answer is yes, it is always there, the interference is most probably internal. When we think of the most likely sources of external interference it becomes obvious that they are not always present. An amateur radio transmitter, as described in the previous example, cannot be on the air continuously. Similarly, diathermy equipment or automobile ignition noise is not constantly present. Even the arcing from a streetlight, a neon sign, or fluorescent lights is not a constant source of interference. The internal arcing of a power transformer mounted on a nearby utility pole is the exception to this rule. But even here, when the defective equipment is removed from the customer to the repairman's shop, the interference will be eliminated.

Here are some quick ways to check and make sure that the interference originates outside the equipment:

If an antenna is used, as in any type of receiving equipment, short the antenna at the terminals, or close to the actual receiver input and observe if the interference is still there. Of course, you will lose the desired signal but, if the interference signal is there by itself, the probability of it originating within the equipment is much more likely.

With the antenna shorted, rotate the equipment in various directions to see if the interference becomes weaker or stronger. If it does, it is obviously external and is being picked up by the cabinet or chassis or some of the internal wiring.

As a third, simple step, move the equipment to a shielded area, such as a concrete basement or a space between concrete walls. This should greatly reduce the pickup due to an external interference signal.

If the interference seems to come in through the AC power line, it may help to put a 0.1 MFD capacitor across the AC power line where it enters the equipment. Be sure that this capacitor has a voltage rating of at least 400 v DC.

Finally, removing the equipment to another location, such as your service shop, should definitely decide whether or not external interference or internal interference causes the apparent defect.

7.3 HOW TO ISOLATE INTERNAL INTERFERENCE

Once it is clearly established that the interference originates within the equipment, it is necessary to determine what kind of interference it is and what causes the appearance of this defect at this time. The best approach to answer these questions is to resort to the symptom-function technique of troubleshooting, which we have explained in detail in the beginning of this book. As you remember, the first step in the symptom-function theory of troubleshooting is to carefully observe the symptoms. The second step is to analyze the functional portions of the equipment and determine which ones could cause defects similar to the observed symptoms. To illustrate how this is done, the following is an example of a type of trouble that frequently vexes the most experienced troubleshooters.

The block diagram in Figure 7-5 shows the functional blocks of a color TV set. Figure 7-6 shows a photograph of the screen of the same color

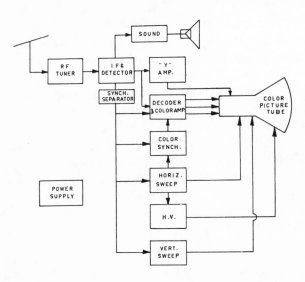

From *Color TV Servicing*, 2nd edition, Walter H. Buchsbaum, ©1968, Prentice-Hall, Inc. p. 38.

FIGURE 7-5
BLOCK DIAGRAM OF COLOR TV SET

From *Color TV Servicing*, 2nd edition, Walter H. Buchsbaum, ©1968, Prentice-Hall, Inc., p. 259.

FIGURE 7-6
HORIZONTAL BARS ON THE SCREEN

TV set with horizontal bars, apparently fixed in the center. The symptom observed is that the screen is darker and lighter along horizontal lines. We can quickly verify that this is not due to outside interference by switching channels and by disconnecting the antenna. As we can see from the photograph in Figure 7-6, a blank screen shows the defect even more. Look at the functional blocks in Figure 7-5. Could this symptom be caused by the audio section? Could it be caused by the horizontal? The color synchronization, the color demodulator, or the video amplifier section? A little bit of deduction and understanding of how a TV receiver works will lead us to suspect immediately that the horizontal bar is synchronized with the vertical scan and this leads us to believe that a 60 Hz or 120 Hz interference is causing it. TV technicians know that it takes 1/60th of a second to scan a field from top to bottom. Suppose that the vertical sawtooth signal would appear in the video. Depending on the polarity, either the top or the bottom of the picture would tend to be much darker than its opposite end.

Now look at the photograph in Figure 7-6. Clearly this is not the case. Referring to Figure 7-7 we can see a sinewave superimposed on the video signal in such a way that two peaks will occur between the beginning and the end of the vertical sweep. Another look at Figure 7-7 convinces us at once that the interfering frequency is 120 Hz.

146

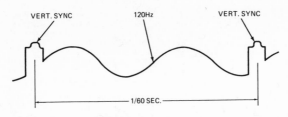

FIGURE 7-7
120 Hz HUM LOCKED IN WITH VERTICAL SYNC

Now look at the block diagram in Figure 7-5 and you will see that the only source of 120 Hz is the full-wave rectifier of the power supply. It is true that the 120 Hz is rectified and filtered, but if the filtering for some reason is not as effective as it should be, some 120 Hz ripple may get into one of the amplifiers along the way and that may be just sufficient to cause the trouble.

Having identified the source of interference as 120 Hz, it still remains to correct the problem. The first approach is to check and possibly replace the filter capacitors. In many TV receivers, different power supply filters provide the DC voltage for the video amplifier, the IF amplifier, or the tuner. A little 120 Hz hum reaching the tuner or the IF amplifier can cause far more problems than if it reaches the video amplifier. This is a consideration in deciding where to look for a defective filter. If we simply ground the input to the video amplifier and the interference disappears, chances are it entered before the video amplifier, probably in the IF stages.

In addition to 60 Hz or 120 Hz hum interference, internal interference can often be caused by a signal from one circuit being superimposed onto another. The usual cause for this is that the B+ voltage is not properly isolated. Figure 7-8 shows a typical case of the DC isolating networks for the audio circuitry and for the vertical output amplifier of the same TV receiver used in the previous example. If C1 were to open, or if its ground were to be a high resistance or intermittent ground, the heavy ripple from the vertical output amplifier would get back to the power supply and could be applied to the input stage of the audio circuit. This would cause a humming or buzzing sound as interference.

This same type of internal interference between circuits can occur easily in digital circuitry where one clock signal or pulse train can

147

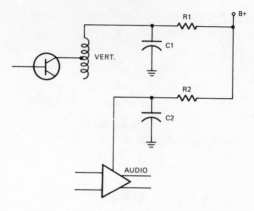

FIGURE 7-8
B+ ISOLATING NETWORKS IN TV RECEIVERS

interfere with that of a different circuit, causing wrong triggering and false results. Unless you are willing to simply replace bypass capacitors and hope for the best, a good oscilloscope is needed to find this kind of trouble and track it down to the defective component. Here, the signal-tracing method, described in more detail in the beginning of this book, is superior to the symptom-function technique. First, however, you have to use the symptom-function technique until you know that interference of superimposition of signals is taking place.

Probably the most difficult problem is internal oscillation and when you run up against that, expect a lengthy troubleshooting job. Practically any high-gain amplifier will oscillate if even the slightest bit of positive feedback is provided. This means that one of the bypass capacitors, negative feedback, or stabilization networks has become defective. If the neutralizing capacitor in a transistor IF stage either becomes open circuit or changes value drastically, chances are that stage will oscillate.

The mode of oscillation of high-gain amplifiers is predictable only by complex calculations and, for our purposes, we only need to know that instead of amplifying, the stage oscillates. The symptom-function technique is often inadequate to determine that self-oscillation has taken place and it certainly cannot tell us where it is coming from. Even in a simple hi-fi stereo set, a typical howl cannot easily be traced to a single stage unless we do some troubleshooting first.

In vacuum tube equipment, the preferred method of troubleshooting this kind of interference is to pull each tube, in succession, until you reach the one tube, or group of tubes, that stops the oscillation. In transistor circuits, and in integrated circuit equipment, this is much more difficult. One method is to suspect every tuned amplifier or other high-gain stage. To check whether or not it oscillates, ground the input to the stage, apply extra bypass capacitors to all points that are normally bypassed, and hope for the best. It is also a good idea, similar to the process described in chapter 6 for intermittents, to check all ground circuits, to solder and resolder all ground connections, and to make sure all internal shields are in place and all coax shields are properly grounded. We told you that internal oscillation was a difficult kind of defect. It is and we hope you don't encounter it often.

7.4 HOW TO ISOLATE EXTERNAL INTERFERENCE

In paragraph 7.2 we have shown you how to separate the internal from the external interference defect. Once we know that the interference comes from the outside, we still have the problem of keeping the interference signal out of the equipment.

First, let us deal with receiving equipment in which the interfering signal is in the same frequency band as the desired signal. In a typical Citizen's Band (CB) receiver, the receiving antenna and the transmission line will, of course, pick up the desired as well as the undesired signal. In mobile CB operations, the desired signal may come from any direction and therefore the receiving antenna must be omnidirectional, which is quite different from a TV receiver situation where the antenna points toward the source of the desired signal. If the interfering signal is in the same frequency band as the desired signal and if it also comes from the same direction, there is no way in which the desired and undesired signals can be separated. In the case of the CB equipment, the only solution is to go to another CB channel. In the case of the TV receiver, the set owner has to accept the fact that reception on at least that one channel is full of interference whenever the interference goes on.

While we can't do anything about this type of interference from the technical side, we often can appeal to the owner of the interfering device, such as the local power company, the owner of a neon sign, a diathermy machine, or a heat-sealing machine, to reduce the interference by proper shielding and other measures. If we cannot readily locate the owner, an appeal to the regional office of the Federal Communications Commission (FCC) should result in some action. Most of these interference sources are supposed to be properly filtered and shielded and, if they are not, their owner is probably violating FCC requirements.

Very often the interfering signal comes in through the antenna or transmission line but is not at the same frequency as the desired signal. Often this interfering signal is within the IF band of the receiver or, more rarely, it directly falls into the video or audio band. In these cases it is possible, by putting appropriate bypass filters at the antenna input, to reduce or eliminate the unwanted signal. Some simple filtering methods are described at the end of this chapter. Many types of radiated external interference are picked up by wires, components, or even the chassis itself. If we can identify the source of the interfering signal and have the trouble cleaned up at the source, as mentioned above, that is the best solution. Frequently this is not possible. In those cases we may have to shield and isolate the equipment or portions of it. In a later paragraph we describe some simple shielding methods in more detail.

In many types of industrial electronics installations, large AC fields exist, usually due to power transformers or rotating machinery that cannot be avoided. The type of electronic equipment usually used in such environments is properly shielded to begin with. The shielding, however, is only effective against electrostatic fields. Electromagnetic fields induced into the chassis and wiring magnetically can cause considerable interference. Magnetic shielding requires special materials such as mu-metal and even this is only of limited value. Since magnetic fields and magnetic pickup are very sensitive to the direction, however, it is often possible to rotate the electronic equipment in such a way that minimum magnetic pickup occurs.

Filtering and shielding are only effective if the filters and shields are properly grounded. In home-type equipment, the electronic circuitry is usually connected to a chassis ground, which is also the ground of the power supply. Where a two-wire AC cord is used, however, the actual grounding of the equipment to the house ground or to the power company ground is uncertain. Reversing the AC plug may improve the ground. Three-wire AC line cords, which contain a separate grounding pin, are much more effective in connecting the chassis ground to the power line ground. This works only, however, if the power line ground is indeed a good ground and does not have a long return path to ground. Some service technicians believe that they can simply take a shielded braid, wrap it around a nearby water or gas pipe, and connect it to the chassis. In those cases where an isolating transformer is used, the device that separates the power line from the chassis, this may work. When a power transformer is not used, and that is the case in a large number, if not the majority of home appliances, this type of chassis ground can result in short circuits. As you can see, there is a great deal to know about grounding.

7.5 HOW TO IMPROVE GROUNDING

As mentioned above, and as will be explained again in the next paragraph, the type of external interference that is picked up by the wiring and chassis can often be reduced considerably by shielding or by using the existing chassis as a shield, if that shield is properly grounded. In industrial installations special grounding connections are usually supplied for electronic equipment that is designed for this application. Very often, the installation instructions for industrial equipment also specify the type of ground that is required. One type of ground is the so-called "earth" ground, that is, a water pipe or a ground rod that represents a low resistance to the earth itself. In radio transmitters, for example, it is often the practice to bury metal cable or extensive pipes in the ground to assure sufficiently good connection to the earth itself. Another type of ground that is frequently used in industrial installations is the so-called power line ground, which corresponds to the neutral point of a Wye-connected three-phase AC power system.

In consumer electronics, computers, and other areas where the standard industrial grounding practices are not followed, the most frequent type of ground is a connection to a water pipe or some other "earth" ground. If you want to connect a hi-fi amplifier, a TV set, or some other home-type equipment to a suitable ground, such as a water pipe, you must first be sure that the chassis or metal shield is not connected in any way to the AC power line. If the equipment uses a power transformer, chances are that the required isolation is provided.

151

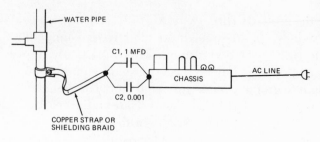

FIGURE 7-9
SAFE CHASSIS GROUNDING

To be absolutely sure that a short circuit does not occur, the safe chassis grounding scheme illustrated in Figure 7-9 is recommended. In this scheme a copper braid, such as shielding braid or a soft copper strip, is connected between a water pipe, gas pipe, etc., and the chassis, isolated by two capacitors. One capacitor should be large enough to pass even low-frequency, 60 Hz signals to ground while the second capacitor is intended for high-frequency signals. A 1 MFD capacitor is usually a foil or paper-wound capacitor that has considerable inductance at higher frequencies. C2 in Figure 7-9 is a .001 MFD capacitor, either of the mica or ceramic type, which has a minimum of inductance.

A word about the connection to the water pipe. To make a good ground it is not sufficient to have a single screw point, which may be pressing against the pipe as part of a galvanized iron strap. The best kind of a ground is against a copper or brass water pipe in which the copper or brass has been sanded down so that the ground strap will make good contact all around the pipe. The whole value of the ground strap can be lost if a high-resistance connection exists between the copper and the water pipe itself. Iron drain pipes or iron gas pipes are not very satisfactory as grounds, particularly for high-frequency use.

If you have been at all involved in shielding and grounding problems, you have heard about so-called "ground" currents. Figure 7-10 illustrates a typical instance of how ground currents can be caused in such a simple thing as a modular hi-fi system. Suppose that the power supply is located separately; chassis 1 contains the tuner and one amplifier, and chassis 2 contains the other amplifier. Suppose a coax cable exists between the two chassis and suppose that both sides of the coax shield are properly soldered to each chassis. Because chassis

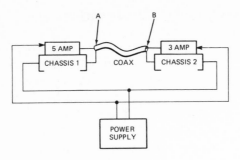

FIGURE 7-10
HOW GROUND CURRENTS ARE CAUSED

1 consumes 2 amperes of B+ current more than chassis 2, some of this current will flow through the coax from one chassis to the other. Even though the coax shield is of a relatively low resistance, the current flowing can easily set up a 1- or 2-volt differential between the two chassis. The power required by both chassis will vary according to the audio output, that is, the music power of each of the two stereo channels. In other words, a varying signal can occur across points A and B in Figure 7-10 and this, in effect, causes a potential difference between chassis 1 and 2, which can give rise to considerable interference. The solution to this type of problem is to provide a good ground return to the power supply and, possibly, isolate the two chassis as far as DC or the low-frequency power supply AC is concerned.

Other types of ground loops occur in high-power, high-frequency RF equipment. This type of signal travels almost exclusively on the surface of the ground plane, copper, brass, etc., and differentials in current between two separate points on a chassis, for example, can cause ground loops. In these cases, just as in the ground currents illustrated in Figure 7-10, separation of the ground loops and recombination to a single point is the usual solution.

One last word about grounding. Never use thin wire or any sort of makeshift arrangement. If the ground is supposed to do something, it must be a very low-resistance path to the earth or to the power line ground. Copper straps or heavy shielding braid are the best connections. Even these should not be longer than 36 inches or so. The connection at either end must be a low-resistance, solid connection that will not become a high-resistance connection due to corrosion in a short time.

7.6 SOME SIMPLE SHIELDING METHODS

The original equipment designer usually provides all of the shielding necessary to avoid internal interference. Occasionally the electronics troubleshooter has to improve on the design or else, because of some external interference, additional shielding has to be provided. The type of shielding that we can add to existing electronic equipment is usually effective only against RF external or internal pickup and does not work at all against low-frequency magnetic pickup.

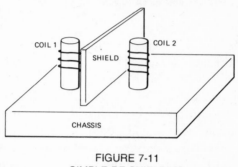

FIGURE 7-11
SIMPLE RF SHIELDING

Figure 7-11 illustrates the principle of RF shielding. Two coils are isolated by a shield placed between them. In order for the shield to be effective, it must be grounded well and it must be large enough to shield each coil against the fields from the other. As a rough rule of thumb, you can be sure a straight shield will work if each of its dimensions is at least 1½ times as large as the component or circuit that you want to shield. For best effect, the shield should be placed halfway between the two sources of trouble. If you place the shield too close to the coil or other component that you want to protect, you may interfere with the operation of that component. In the case of the coil in Figure 7-11, placing the shield too close to either coil would tend to reduce the Q of that coil.

In many cases it is possible to purchase shielding cans that readily fit over coils or circuits that are particularly susceptible to interference. In other cases it may be simpler to go over your collection of junk and look for some shielding cans, such as are used in the IF stages of TV sets and old-fashioned radio receivers. It is essential, no matter what

154

kind of shield you use, that a good, low-resistance ground be provided for the shield.

On some rare occasions we encounter situations where outside interference is very strong and where we cannot get at the source of that interference. As a last resort we have used shielding boxes, as illustrated in Figure 7-12, to eliminate the effects of outside interference. This simple type of shielding box can be constructed from a sheet of copper or aluminum screening. You should not use copper screening with an aluminum base or vice versa because the dissimilarity of the two metals will result in a poor ground connection. Figure 7-12 shows that the equipment to be shielded is mounted on insulators. This means that, in effect, a capacitive isolation exists between the shield box and the equipment and this adds to the effect of the shielding. Note also that shielding is provided for the length of the AC cord going to the equipment, such as BX cable, and the AC power line itself is bypassed, for RF, by a capacitor to the shielding box.

Many RF devices and systems are tested in the laboratory inside a screened room. This is a chamber, carefully constructed of an outer and inner shielding material, with the two insulated from each other with an specially filtered AC power inputs. This type of shielded room can provide rejection of unwanted signals in the order of 100 db or more. If you ever have to construct a shield box, something like the one shown in Figure 7-12, you will be doing well if you can get an attenuation of 70 or 80 db. For almost all types of interference, however, that is plenty.

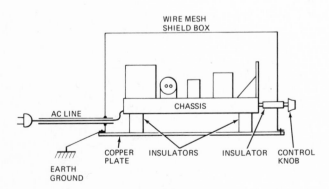

FIGURE 7-12
RF SHIELD BOX

7.7 SOME SIMPLE FILTERING METHODS

When you run into an interference problem that can best be solved by installing a filter, you may want to try to purchase a suitable filter or else build it yourself. If the interfering signal that you want to keep out is higher in frequency than the signals that you want to leave in, such as amateur radio transmissions being picked up in the video portion of a TV set, the filter that you want to install is a low-pass type, such as that shown in Figure 7-13. This type of low-pass filter is usually called pi, after the Greek letter that it resembles. The same type of filter can also be designed in the form of a T where the vertical portion of the T consists of one capacitor and the top of the T of two inductors. The formulas for the values of the coil and capacitors and the characteristic, or load, impedance of the filter, Z, are shown in the illustration. The example worked out in Figure 7-13 assumes a load impedance of 75 ohms, typical of the coax cable used in TV sets and a cut-off frequency of 5 MHz. At the cut-off frequency the response of the filter is 3 db down from the flat portion, which, in this case, is below the cut-off frequency.

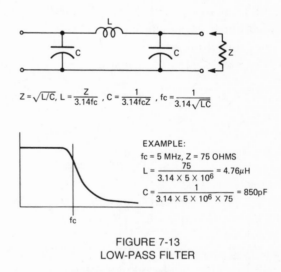

$$Z = \sqrt{L/C}, \ L = \frac{Z}{3.14fc}, \ C = \frac{1}{3.14fcZ}, \ fc = \frac{1}{3.14\sqrt{LC}}$$

EXAMPLE:
fc = 5 MHz, Z = 75 OHMS

$$L = \frac{75}{3.14 \times 5 \times 10^6} = 4.76\mu H$$

$$C = \frac{1}{3.14 \times 5 \times 10^6 \times 75} = 850pF$$

FIGURE 7-13
LOW-PASS FILTER

Figure 7-13, except that the shunt elements would be coils and the If the interfering frequency is lower than the signal that you want to pass, you will use a high-pass filter. Just to illustrate how a T-type high-pass filter could be constructed, Figure 7-14 shows this confguraton. A pi filter could be constructed similar to that shown in

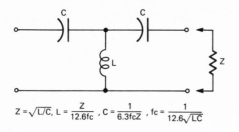

$$Z = \sqrt{L/C}, \; L = \frac{Z}{12.6fc}, \; C = \frac{1}{6.3fcZ}, \; fc = \frac{1}{12.6\sqrt{LC}}$$

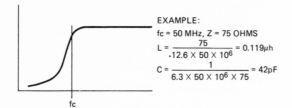

EXAMPLE:
fc = 50 MHz, Z = 75 OHMS

$$L = \frac{75}{.12.6 \times 50 \times 10^6} = 0.119\mu h$$

$$C = \frac{1}{6.3 \times 50 \times 10^6 \times 75} = 42pF$$

FIGURE 7-14
HIGH-PASS FILTER

series element a single capacitor. In the example worked out in Figure 7-14, we have again assumed that the characteristic impedance is 75 ohms but this time the cut-off frequency is assumed to be 50 MHz. This would be the case where we want to reject all lower frequency interfering signals that could come in on a TV antenna coax transmission line. In a high-pass filter the frequencies below the cut-off frequencies are suppressed and the cut-off frequency itself is 3 db down from the flat response at the higher frequency end.

The design of more complicated filters requires special knowledge, calculations, and the ability to select and build specific components. If you simply want to reject one particular interfering frequency, it is often possible to use the simplest approach, a shunt or series resonant circuit. In Figure 7-15 we show you a parallel resonant circuit consisting of L_p and C_p and its equivalent, a series resonant circuit that is shunted across the load impedance. The parallel resonant circuit has very high impedance at the resonant frequency. The series resonant circuit has a very low impedance. The values for L_p, C_p, L_s, and C_s can be calculated from the resonance formula shown in Figure 7-15. You may find it useful to refer back to this when you run into a situation where a single frequency is interfering with the desired signal and when you want to "trap out" this particular frequency from the circuit. Depending on the type of circuit it is, you can use the

157

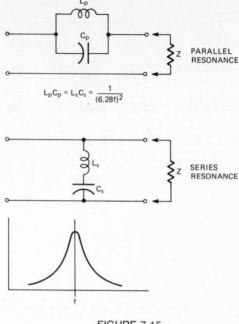

FIGURE 7-15
SINGLE-FREQUENCY FILTERING

parallel resonant circuit or the series resonant circuit, or a combination of both, to attenuate the undesired signal. The higher the Q of the coil and the capacitor are in this approach, the more effective it will be and the narrower will be the bandwidth over which the rejection of attenuation characteristic of this resonant circuit will act on the interfering signal.

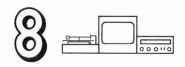

Troubleshooting
TV and Hi-Fi Equipment

8.1 WHAT YOU NEED TO KNOW ABOUT THE EQUIPMENT

Home entertainment equipment, such as TV receivers and hi-fi systems, differs from most other electronic equipment because it is mass produced and designed to meet a price. In recent years much of the hi-fi equipment and almost all black-and-white TV receivers have been produced in various places in Asia, including Singapore, Taiwan, and Korea. As a result, many of the components are not labelled or numbered, and sometimes don't even look like their American counterparts. The overall packaging of some of this equipment is also different from what we are used to. Most American equipment is mounted on some form of chassis and, by removing the proper screws, the cover or case can be slipped off and the chassis is available for servicing. In some foreign equipment it is sometimes difficult to tell where the chassis ends and the cover portion begins. Of course, foreign-made TV and hi-fi equipment operates on the same electronic principles and with the proper manufacturer's manual on hand we can troubleshoot it just like American equipment.

What we basically need to know about the TV and hi-fi equipment we are going to troubleshoot is:

1. How does it work?
2. How can we test it?

The answer to the first question requires a knowledge and understanding of TV receivers and of hi-fi audio systems. If you want to really delve into it, you can get any one of a number of very good textbooks that contain all of the operating details of these equipments. For color TV servicing read *Color TV Servicing by* W. H. Buchsbaum, published by Prentice-Hall, Inc., and for hi-fi servicing we recommend the *Audio Systems Handbook* by TAB Books.

For those who want to get some idea of how TV and hi-fi equipment works, we present basic block diagrams and brief explanations in the following paragraphs.

Monochrome TV receiver

The block diagram in Figure 8-1 shows all of the essential functions of a monochrome TV receiver. Starting at the upper left-hand corner, a high-pass filter (H.P.F.) brings the signal from the TV antenna to the VHF tuner, which contains, in this case, three transistors. UHF channels are received over the UHF tuner. The VHF tuner acts just like one more IF stage for the output of the UHF tuner. The VHF tuner gain is controlled by the AGC amplifier. At the VHF tuner the RF signals are amplified, selected, and converted down to the intermediate frequency (IF) from about 41 to 45.4 MHz. The video IF amplifier (VIF AMP) uses three transistors and a number of tuned circuits to provide the required gain over the 3.8 MHz bandwidth. At the video detector a diode is used as peak detector to demodulate the actual video signal. The video drive and the video output amplifier then amplify the video signal further until it is applied to the picture tube.

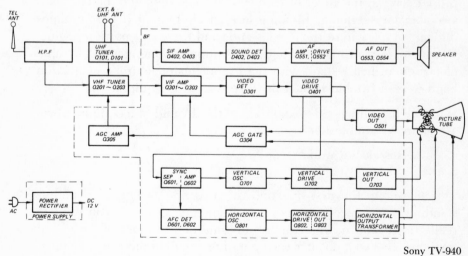

Sony TV-940

FIGURE 8-1
MONOCHROME TV BLOCK DIAGRAM

At the video detector output the sound IF amplifier (SIF AMP) uses two transistors to amplify the 4.5 MHz intercarrier sound signal. This signal is applied to the dual diode FM sound detector section where the audio signal is demodulated. This audio signal goes to the audio driver and output stage and then drives the speaker.

The video driver stage also supplies video signals to the AGC gate and to the synchronizing separator and amplifier portion. At that point the vertical synchronizing pulses are separated to drive the vertical oscillator and the horizontal synchronizing pulses go to the automatic frequency control (AFC) and detector circuit. Although the block diagram does not show it, the AFC portion compares the incoming horizontal pulses with those generated in the receiver and develops an error voltage to keep the horizontal oscillator properly synchronized with the transmitter. A two-stage horizontal drive and output amplifier provides considerable power to the horizontal output transformer. The high voltage for the picture tube is obtained by rectifying the horizontal flyback pulses in the output transformer.

The only remaining block is the power supply, which rectifies and adjusts the voltage to the levels required in the receiver. Later in this chapter we will refer back to this block diagram and its explanation again.

Color TV receiver block diagram

Figure 8-2 shows the detailed functional diagram of a typical modern color TV receiver. In comparing this illustration with the block diagram of the monochrome receiver we can immediately spot the common functions. Clearly, the audio portion is essentially the same for both. Similarly, the tuner and video IF sections, together with the AGC functions, the vertical and horizontal synchronizing circuits, and power supply must be basically the same. All these functions are common to a monochrome and to a color TV receiver.

Starting at the picture tube, we see that there are three separate video sections labelled red, blue, and green. Each section contains a demodulator, a driver, and an output stage. The last stage has a common input from the blanker, which receives pulse signals from the horizontal and from the vertical deflection sections. The blanker cuts off all three video signals during the retrace time. The three driver stages (red, blue, and green) have a common input from the

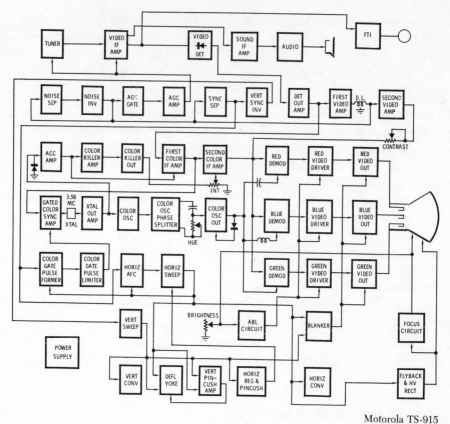

FIGURE 8-2
COLOR TV BLOCK DIAGRAM

Motorola TS-915

automatic brightness level (ABL) circuit. That circuit is controlled by the setting of the brightness control. In other words, the gain of the three driver stages is controlled by an automatic level circuit.

To understand how the color TV receiver produces a composite color picture from the red, green, and blue video signal requires an understanding of the basics of colorimetry. For our purposes, it is sufficient to know that the color information is transmitted to each color TV receiver by means of a 3.58 MHz color subcarrier, interleaved within the video bandwidth. This color subcarrier is both amplitude and phase-modulated. In the block diagram in Figure 8-2, the first and second color IF amplifiers provide an approximately 2 MHz wide band pass amplification for the 3.59 MHz color subcarrier and its sidebands. The same color subcarrier is applied to all three color

162

demodulators. What differs in each of the demodulators is the color synchronizing signal.

At the transmitter an 8-cycle burst of 3.58 MHz reference signal, the fixed reference phase that permits phase demodulation of the subcarrier, is transmitted with each horizontal sync pulse. In the block diagram in Figure 8-2, the gated color sync amplifier removes this 8-cycle burst from the horizontal sync pulse and excites the 3.58 MHz crystal. The crystal output amplifier controls the color oscillator and this, in turn, feeds the color oscillator phase splitter. At that point, the hue control determines the exact reference phase. The color oscillator output stage drives two phase-shifting networks, one indicated by a capacitor and one indicated by an inductor. This means that if the reference signal from the color oscillator output stage is applied to the green demodulator, it will be plus or minus 90 degrees out of phase at the red and blue demodulators respectively. The phase differences account for the different results of the red, blue, and green color demodulators. Each demodulator constantly compares the 3.58 MHz color subcarrier against the reference signal and thereby produces a video signal that corresponds to the red, blue, or green component of the original color picture.

Because of the alignment problems of the three electron beams in a shadow mask tube, a number of correcting circuits are required. These include the horizontal and vertical convergence circuits and the horizontal and vertical pin cushion and regulating circuits. These circuits, generally passive circuits with adjustments for each color, are used to make sure that the three color electron beams are properly converged all over the color dot screen.

AM-FM tuner

The block diagram in Figure 8-3 is considerably simpler than those for the TV receivers. In effect, we see two separate receivers, one for the AM and one for the FM channels. The FM portion contains an RF amplifier, mixer, and oscillator, all of which are tuned by the front panel channel selection control. The IF amplifier is the same for FM as it is for AM, although different frequencies are employed. The FM signals are amplified at 10.7 MHz, while the AM signals are amplified at 455 KHz. At the output, the FM limiter and detector changes the 10.7 MHz IF signal into the detected audio signals. On those channels having stereo multiplex, the MUX decoder section decodes the

163

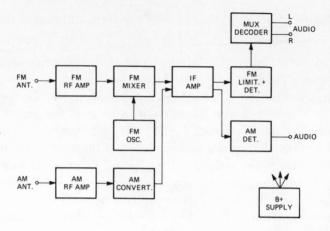

FIGURE 8-3
AM/FM TUNER BLOCK DIAGRAM

stereo carrier and the result is two separate audio channels, one left and one right.

The AM section resembles, in every respect, a typical AM radio. The RF amplifier and the converter really perform the same heterodyning functions as in the case of the FM receiver. We use the term "converter" because the same tube or transistor stage is used for mixing and as the local oscillator frequency. The IF amplifier is common and the AM detector is, in most cases, a simple diode peak detector. Since both AM and FM cannot be used at the same time, the power supply is switched between those two respective sections.

Stereo amplifier

For our discussion, only one of the two identical stereo amplifiers has to be considered. If trouble occurs, we can always tell in which amplifier it is because we can turn either one off and then listen to the other. Figure 8-4 shows a typical stereo amplifier consisting of two preamplifiers and one power output stage. The first preamplifier is used only for preamplifying the phono output and consists of a single IC. The second preamplifier, which will amplify the output of the tuner, the tape, the auxillary or other inputs, consists of an IC and a transistor driver stage. The function selection switch, the volume control, and usually the tone controls as well, are always located before the power amplifier stage. This stage, in Figure 8-4, is shown to consist of ten transistors. Different output points, at different

164

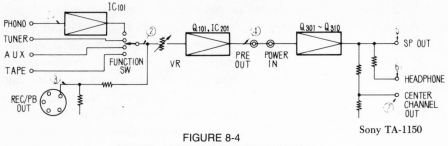

FIGURE 8-4
STEREO AMPLIFIER BLOCK DIAGRAM

Sony TA-1150

levels, provide the audio signal to the speaker, to headphones, and to a center channel output, a network that can be used to provide output to a third speaker.

The sole function of all of the stereo amplifier blocks is to amplify the signal. For this reason, the main criteria as to whether the unit functions properly or not is a measurement of the signal level. As shown in Figure 8-5, a level diagram can be plotted to show the relative signal levels, both in db above or below zero dbm, and in terms of millivolts. When we apply the information from the level diagram to the block diagram of the stereo amplifier as shown in Figure 8-4, we can easily

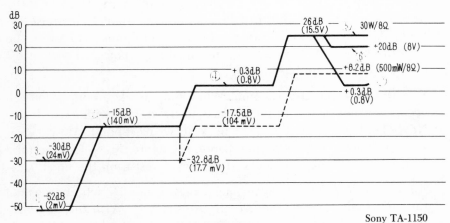

Sony TA-1150

FIGURE 8-5
AUDIO LEVEL CHART FOR AMPLIFIER IN FIGURE 8-4

165

determine where the defect occurs. With a signal input of 140 millivolts at point 2, for example, we should get 0.8 volts at point 4 with the volume control set to maximum. If we get less gain than that, something is wrong in the preamplifier. Of course, frequency response and linearity of amplification, freedom from noise, hum, and other interference are also important parameters in the operation of stereo amplifiers. From a troubleshooting point of view, however, a loss of amplification is involved in most other defects.

The more you know about the equipment, the easier it is to troubleshoot. If you know as much about it as we have covered in this chapter so far, you should be able to troubleshoot most defects in TV receivers and hi-fi equipment. As you work at it, you will gain additional knowledge and insights, but remember—using the right technique is half the game.

8.2 HOW TO USE THE SYMPTOM-FUNCTION TECHNIQUE

CAUTION—DANGER—CAUTION

This warning applies to all service work on hi-fi equipment and even more so for TV work. One danger comes from the AC power line if the chassis is not isolated through a power transformer. One hundred seventeen volts AC can cause deadly shocks. *Be careful.*

TV receivers use high voltages at the picture tube, ranging from 10,000 to 20,000 volts. While the current is low, nasty burns and the possibility of injuries due to falls make it important to exercise extreme caution. The high-voltage capacitor can retain a charge, so be sure to ground the high-voltage terminal *after power is turned off* before you start troubleshooting.

Picture tubes contain a vacuum. When they break, an implosion can occur, causing glass and metal parts to fly around with great force. Never hold a picture tube at the neck, its weakest point. Always wear safety goggles and gloves when handling picture tubes. Lay the tube face down on a soft, padded surface; never rest it on the neck.

A final word of caution concerning x-rays. Color TV sets, because of the very high voltages used, emit "soft" x-rays. With the manufacturer-supplied shielding, there is no danger, but when shielding is removed it is possible that persistent exposure could cause x-ray damage, particularly to the eyes. Always operate TV sets with all high-voltage shielding in place.

Every TV receiver contains an audio section. There is only a difference in quality and in the sophistication between the audio section of a TV receiver and the hi-fi or stereo audio amplifier. To apply the symptom-function technique to either type of audio system requires only an attentive ear, some quiet, and a little imagination. The most frequent defect is a total lack of sound and this means that somewhere along the way either an amplifier is dead, a connection is open, shorted, or some other catastrophic failure has occurred. When vacuum tubes are involved, the simplest approach is to change tubes. With solid-state devices, however, it is necessary to locate the defective stage. First, we check that all of the plugs and jacks, sockets and tie-points form good connections. Next, we check to make sure that there is B plus. The third check is to temporarily ground signal points, either going backwards from the loudspeaker output or coming toward the loudspeaker from the low audio input end. In either way we will reach the point where a click is no longer heard or a click is first heard. As long as we hear a click, we know that there is some amplification between the point we shorted and the loudspeaker. Between that shorted point and the shorted point further away from the loudspeaker at which we hear no click is the most likely defect.

If the defect is a distortion, weak sound, noise, interference, etc., we have to listen to it carefully, adjust the volume and tone controls, and figure out from these factors what could really be wrong. A noisy control is obvious because we hear the noise as we rotate the knob. Sixty Hz hum or 120 Hz hum are also relatively easy to recognize, particularly if they are always present. These symptoms are due to excessive ripple in the power supply and this usually points to an open filter capacitor. Distortion of the sound occurs in two different ways. One type of distortion occurs only when the volume is turned up all the way. This symptom indicates that the speech is being clipped or distorted in some amplifier after the volume control, due to a nonlinearity in that amplifier. If distortion occurs regardless of the setting of the volume control, it is most likely due to a leaky coupling capacitor or an open bypass capacitor somewhere in the audio amplifier circuit. The symptom-function technique usually allows us to focus our detailed troubleshooting, signal-tracing, or resistance-voltage measurements to a few components.

167

In AM-FM tuners the symptom-function technique is usually limited to determing whether there is an overall loss of amplification, such as is the case when only a few stations can be received, or whether either the AM or the FM portion does not work at all. For AM-FM tuners the signal-tracing technique is most useful.

Other audio equipment, such as record players and tape recorders, contain only a small portion of electronic circuits and these are essentially audio. Most of the troubles in those equipments is mechanical or electromechanical. Here the symptom-function technique can be of great help, provided we understand the mechanical operation of the device. In the case of the tape recorder, for example, we once spent an entire afternoon tracing down the trouble until we found that the drive belt that controlled the tension on the take-up reel had stretched and was slipping. Troubleshooters who are more familiar with this type of equipment will automatically look for the obvious mechanical defects first. On record players it is possible to examine the needle under a microscope to determine whether it is defective or not. It is not possible to examine the cartridge in a similar way and, if distortion originates in the record player, it can be either the cartridge, the stylus, or the record itself. A test record or one known to be good should be used to make sure that it is not the record itself. The cartridge and stylus can only be replaced.

In monochrome TV receivers the symptom-function technique should help you to isolate trouble to at least one or two stages. When we look at the block diagram in Figure 8-1 and Figure 8-6, table of TV trouble symptoms and the receiver functions that can cause them, it becomes easy to see the connection between the observed symptom and the function. Assume, for example, that the video drive transistor is defective. This means there will be no video on the picture tube and therefore no picture. Does this also mean that there will be no raster? Of course not, because the raster is produced by the vertical and horizontal deflection circuits and requires the presence of high voltage, which is obtained from the horizontal output transformer. In other words, the picture tube will light up, but will show a blank screen. Should the audio be affected? Certainly not, because the audio signal is taken off before the video drive circuit. To sum up, then, the fact that there is no picture on the picture tube, but that there is sound and a raster, leads us to suspect either the video drive or the video output stage.

SYMPTOM	FUNCTION
No raster, no sound	Filaments, "B+," or AC line
Sound, but no raster	High voltage, flyback, "B+," or video and brightness control, CRT
Raster, but no sound or picture (no "snow")	Antenna, lead-in, tuner, IF and detector sections
Raster and sound, but no picture	Video amp., contrast control, AGC
Weak, snowy picture; sound may be OK	Antenna, lead-in, tuner, IF detector, video and AGC section
Picture OK, sound weak or distorted	4.5-mc, sound IF, FM detector, audio-output stages
Top or bottom of picture compressed or stretched	Vertical-sweep section; "B+" boost circuit
Bent or twisted pictures; edge tearing	Horiz. osc. or output, sync separator or video amp.
Picture rolls up or down	Vert. osc. or sync separator; video amp.
Picture appears "sliced" or weaves back and forth	Horiz. osc., sync separator, or AGC (if keyed)
Dark horiz. bar in picture on all channels (60 cps in video)	Tuner, IF, or "B+"
Picture fades rapidly on all channels	Antenna, lead-in, tuner, AGC
Picture too narrow	Horiz. output, horiz. osc; damper circuit
Picture too short, top and bottom	Vert. osc., output, or "B+" boost-voltage section
Picture changes size when contrast or brightness is changed	Flyback and H.V. section; picture tube gassy

FIGURE 8-6

TABLE OF MONOCHROME TV SYMPTOMS AND FUNCTIONS

If there were no picture and no sound as well, the logical conclusion would be that one of the stages before the point where the sound IF is taken off is defective. We should now suspect the tuner, the video IF amplifier, and the video detector stage. Of course, a defect in the AGC system could also cut off all signals completely.

If a single horizontal line instead of a raster is shown, the defect is obviously somewhere in the vertical deflection circuits. A vertical line, which would indicate a defect in the horizontal deflection circuits, is extremely rare. It would indicate a defect in the horizontal deflection yoke winding since the rest of the horizontal output section, which also supplies the high voltage, would have to be operating properly.

If the picture continuously moves up or down, the vertical synchronizing section of the circuit from the sync separator to it is suspected. If it is impossible to lock the picture in horizontally, the horizontal AFC portion will, of course, be suspected.

In addition to the many obvious and logical symptom-function relationships, there are also others that are not as clearly defined. Excessive fading of the picture when airplanes fly overhead, for example, could indicate a malfunction in the AGC. It could also indicate that something is wrong either in the antenna or in the transmission line from the antenna. Because a TV receiver is quite a complex piece of electronic equipment, the symptom-function technique can be quite helpful in isolating trouble to a particular section.

The color TV receiver, as evidenced from the block diagram in Figure 8-2, contains all of the functions of the monochrome receiver and, in addition, contains the color-producing functions. When using the symptom-function technique to troubleshoot color TV receivers, the first thing we must do is to mentally separate the symptoms that apply to monochrome from those that apply to color. We can do this by turning the color intensity control down to zero so that the black-and-white picture appears. Now, Figure 8-6 will apply. If the trouble is not part of the black-and-white operation, the table in Figure 8-7 should be consulted, which lists the basic symptoms of color defects and the circuit functions where they can be caused. The symptom-function technique really becomes useful in color TV receivers. Without the symptom-function technique it would be almost impossible to find many defects in color TV receivers.

8.3 HOW TO USE SIGNAL TRACING

TV receivers and hi-fi equipment have the one aspect in common that, essentially, a weak signal is amplified, changed in frequency, demodulated, etc., until it is strong enough to drive a loudspeaker or

SYMPTOM	FUNCTION
No color on any station	Band pass amplifier, color-burst circuit, color-killer circuit.
No color on some stations	Tuner alignment, IF alignment, color-killer level setting.
Weaving bars of red, green, and blue.	Color sync circuit, color-burst circuit.
One primary color (R, G, or B) missing.	Color demodulator, matrix circuit and amplifier, picture tube screen control misadjusted.
All wrong colors (flesh tones, green or purple)	Tint control, or color demodulator color sync circuit.
Weak pale color	Band pass amplifier, video IF section, tuner misaligned.
Excessively strong colors	AGC to tuner or IF section misadjusted, color gain (intensity) control.
Color uneven over screen (Can't get a uniform "grey" picture)	Purity adjustment, degaussing required.
"Color fringing" on either side	Horizontal dynamic convergence.
"Color fringing" at top or bottom	Vertical dynamic convergence.

FIGURE 8-7
TABLE OF COLOR TV SYMPTOMS AND FUNCTIONS

the picture tube. This means that signal tracing can be done very effectively, because a weak signal can be injected, either from a signal generator or from an antenna source, and this signal can then be traced through. If you look back to chapter 1 where we first discussed the signal-tracing technique, you will see that there are two ways of doing it. You can either move the signal source or you can move the indicator, which may be the VTVM, oscilloscope, or, in the case of stereo or TV equipment, even a set of earphones. The approach that you use for a particular type of equipment will depend on what kind of test equipment you have available and, in any event, what is handiest for your particular setup.

In signal tracing audio circuits, whether the audio section in TV or stereo amplifiers, the signal source is usually an audio signal generator capable of providing sinewave signals from about 100 to 15,000 Hz. Standard test frequencies are either 400 or 1,000 Hz. As detector for signal tracing of audio circuits we can use an oscilloscope, which permits us to see any distortion, a VTVM, which allows us to measure gain but does not show up distortion or hum, and finally the human ear, coupled to the system by means of a set of earphones and a high-impedance amplifier. Many technicians who do a lot of audio work keep a hi-fi amplifier and speaker just for the purpose of serving as test-tone indicator. They simply use the high-impedance input to the audio amplifier to connect to the various stages of the audio equipment under test. By adjusting the gain control, which they have previously calibrated, they get a pretty good idea of the stage-by-stage gain in the equipment under test. Instead of a loudspeaker, a set of headphones is often used in locations where other people are also troubleshooting equipment.

For signal tracing TV receivers, different types of signals must be used for different receiver portions. This is why the symptom-function technique is so important in isolating trouble to a particular receiver section. If, for example, we have isolated a defect to the video IF section, then we must have a means of generating the video IF signal. In this case, the most effective way is to apply an oscilloscope to the output of the second detector and put the video IF generator output onto the last IF stage. If a video signal is now apparent on the oscilloscope, the signal generator is moved one stage further away, to the second or first IF stage input, and the amplitude of the generator signal is reduced. In this manner, stage-by-stage, the entire video IF amplifier section can be traced through. Where a sweep frequency signal is not available, the signal generator must be tuned to the different IF frequencies, according to the band-pass requirements described in the manufacturer's literature. This also provides a means of checking the frequencies of the various tuned circuits, both those that amplify and those that reduce, such as traps, particular video IF frequencies.

In monochrome and color TV receivers, if the video signal is available at the second detector, signal tracing can usually be simplified if a good off-the-air signal can be used. Simply tune the receiver to a strong station and then use those video signals, including the color subcarrier, color synchronizing burst, etc., for troubleshooting. For this type of work a good oscilloscope is essential. The oscilloscope must be properly synchronized to display the correct signal. If, for example, we are checking into the color sync section operation, we have to first lock onto the 15,750 Hz horizontal sweep signal. If, for some reason, the internal sync of the scope does not do the job right, you might try to connect the external sync input of the scope to some convenient point of the horizontal flyback circuit. Don't use the high-voltage portion, of course, but some of the points in which a flyback pulse is fed back, say, to the AGC section. In order to show the 3.58 MHz color sync burst, the oscilloscope must have the proper bandwidth and deflection speed.

For signal tracing and alignment of the color demodulator and matrixing circuits, the off-the-air signal is not always easy to use because colors keep changing. If you do a sufficient amount of color TV work, the investment in a color bar generator may be worthwhile. That device provides you with a fixed 3.58 MHz reference signal and with a color subcarrier, the phases of which are shifted, in known steps, to produce standard color bar patterns.

8.4   HOW TO FIND THE DEFECTIVE COMPONENT

In previous chapters we have followed the signal-tracing technique with the voltage-resistance measuring technique to find the defective component and, in some instances, we have used the substitution technique to assure that a part is defective. For TV receivers and hi-fi stereo equipment these two techniques can be combined. Where manufacturer's data containing voltage and resistance values is available, this is a good way to locate a defective component after the suspected circuit has been isolated by means of the symptom-function technique and/or signal tracing. Unfortunately, however, DC measurements often do not reveal that an integrated circuit is defective. Similarly, leaky capacitors are sometimes difficult to find by means of the voltage-resistance technique. Depending on the type of construction that the manufacturer of the equipment has used, it may be easier to simply replace the component.

In the case of audio distortion, where a leaky coupling capacitor is suspected, the simplest approach is to clip off one of the two leads and temporarily solder a replacement capacitor in place. Semiconductor circuits often use polarized electrolytics and this means that we have to make very sure that the polarity of the replacement capacitor is correct. In other circuits, particularly where DIP ICs are used, it is much simpler to measure voltages and resistances than to replace a 14- or 22-pin DIP unit. In any case, with a little experience you will be able to judge whether the voltage-resistance measurement or replacement is faster.

In equipment that uses vacuum tubes or in those portions of the equipment where relatively high power is used, such as the audio output circuits, it is always worthwhile to look for signs of overheating. Charred resistors, melted insulation, discoloration of coating or color code markings—all these are signs that a particular component and often its neighbors have overheated. Simply replacing that component is not likely to cure the problem. Here is where you should know the entire circuit. If necessary, try to trace it out from the PC board and the components but, of course, it is much easier if you have the manufacturer's schematic. Study the circuit carefully to see what could cause overheating. In transistors remember that a change in the transistor characteristics may cause "thermal runaway." Sometimes there are components or resistor networks to prevent just this condition and when they fail thermal runaway will occur.

Before deciding that a defective component is at fault, be sure to look for shorts or poor connections. Refer to chapter 6 for locating intermittent defects. If a transistor, integrated circuit, or tube circuit has been isolated as the cause of the defect, it may be helpful to refer to chapters 3, 4, or 5 respectively for troubleshooting methods of this type of circuit. In any event, once you have located the circuit in which the defect occurs, troubleshooting a TV receiver or a hi-fi amplifier is no different from troubleshooting any type of electronic circuit.

8.5 FREQUENTLY FOUND TROUBLES IN TV RECEIVERS

While we can't predict what kind of troubles you will find in the TV receivers that come your way, we know from our experience and from reports from many other people in the field, that the following troubles occur most frequently.

Horizontal oscillator and flyback

In the block diagram in Figure 8-1 four sections are used to perform

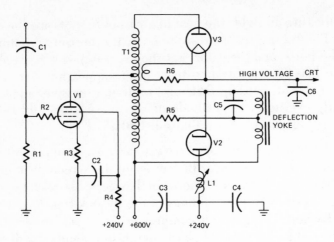

FIGURE 8-8
TUBE-TYPE HORIZONTAL FLYBACK CIRCUIT

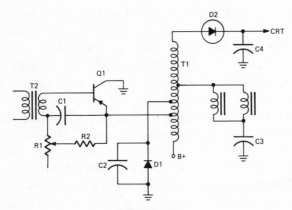

FIGURE 8-9
TRANSISTOR-TYPE HORIZONTAL FLYBACK CIRCUIT

the horizontal deflection and high-voltage generation. Trouble most frequently occurs in the horizontal drive and output section and in the output transformer circuit itself. Figure 8-8 shows the circuit diagram of a typical vacuum-tube type horizontal output and flyback section. Figure 8-9 is the equivalent of this function as used in some transistor TV sets. In vacuum-type sets, the horizontal output tube, V1, is the most frequent source of trouble. If you have no high voltage, that is the first tube to replace. Surprisingly, the high-voltage rectifier, V3, is the least suspect. The damper diode, V2, is usually a very rugged tube but, of course, it can also be defective.

175

If the tubes are all right, the next likely trouble spots are the cathode resistor R3 and the screen resistor R4 of the horizontal output tube V1. If the horizontal oscillator, not shown in Figure 8-8, were defective, the bias on V1, the negative voltage appearing across on R1, would be absent, causing that tube to draw excessive current. This frequently overheats the cathode or screen resistors and they may either open or else change value substantially.

Defects in the deflection yoke are very rare. An open circuit causes a vertical line somewhere near the center of the screen. The high voltage is okay, but there is no horizontal sweep. Short circuits in the deflection yoke or in the windings of the flyback transformer T1 are apparent when the picture is compressed or expanded at either the left or the right side. Open or short circuits in C3 and C4 will result in lack of B+ applied to the output tube V1 and no high voltage. Similarly, if the linearity coil L1 is open, no high voltage will appear.

In the transistor circuit in Figure 8-8, the most likely defective part is the horizontal output transistor Q1, although any of the transistor stages preceding it, such as the oscillator or driver, may also cause loss of high voltage. If the bias adjustment of Q1 is not properly set and if one of the previous stages fails, Q1 may be subject to thermal runaway. In any event, in replacing Q1 be sure to mount the replacement transistor on the heat sink in the prescribed manner. Because the flyback voltage across the damper, D1, is so much lower than it is in the vacuum tube circuit, this diode does not usually cause trouble. The high-voltage rectifier, D2, usually consists either of a series of diodes or else a voltage-multiplier arrangement, which includes series capacitors. Defects in that area are quite common. All of the other troubles that can occur in the vacuum tube horizontal output circuit can also occur in the transistor version.

Vertical sweep circuit

Referring back to the block diagram in Figure 8-1, we see that three stages are used for the vertical deflection: the oscillator, the drive, and the output stage. If any of these three stages becomes defective, there will be no vertical deflection and you will see a horizontal line across the screen. The most frequent vertical trouble is loss of vertical sync. In this defect the picture rolls up or down and adjustment of the vertical hold control does not seem to lock in the signal permanently. This type of trouble can be due to a defective sync separator and

176

amplifier, or to a shorted capacitor in the network connecting that stage with the vertical oscillator. You have adjusted the vertical hold control correctly when the picture rolls from the bottom up and then snaps into place. Switch to different channels to make sure that the vertical hold stays locked in at each channel.

Another frequent trouble in the vertical sweep section is nonlinearity. Either the top or bottom or both appear squeezed or expanded. There are usually two secondary controls, frequently mounted at the rear of the receiver, labelled vertical linearity and height. You have to repeatedly adjust both of these controls to obtain the proper vertical linearity. If you have a crosshatch or a dot generator, you can use this pattern on the screen to adjust the vertical linearity for the optimum setting. If vertical linearity cannot be adjusted properly, the first thing to do in a vacuum tube set is to replace the vertical output stage. In transistor type receivers shorted or leaking capacitors can cause this problem.

Interference

Chapter 7 has dealt extensively with interference defects of all sorts. The most frequent type of interference encountered in TV receivers comes from external sources such as other TV receivers, amateur radio operators, diathermy machines, etc. Before proceeding with some of the techniques described in chapter 7, you should first determine what type of interference it is that the customer complains of. Find out at what times of the day or night it occurs, on what channels it occurs, and any other pertinent data. Detailed instructions are found in chapter 7.

Tuner contacts

The TV tuner or channel selector mechanism involves moving contacts and this is always a source of trouble. In addition to the contacts themselves, all TV receiver tuners have a detent mechanism. This mechanism depends on a spring to lock the channel selector switch into place. With time, this spring weakens and the detent is less certain so that "off-detent" tuning is possible. Tuner defects of this nature are easily recognizable because the picture and the sound changes as you wiggle the channel selector switch. It is relatively simple to clean the contacts and lubricate the entire mechanism. Your electronics distributor carries a whole line of difference spray cans for

177

this purpose. In extreme cases, when the cleaning and lubricating does not help, it may be necessary to disassemble the tuner and carefully bend individual contacts, replace springs, etc. This is generally a job for a specialist and, because it is a time-consuming job, many TV troubleshooters prefer to replace the entire TV tuner.

8.6 FREQUENTLY FOUND TROUBLES IN FM/TV ANTENNA SYSTEMS

Most of the troubles in antenna systems are much more pronounced on TV reception than they are on FM. Of course, when the antenna transmission line is broken, intermittent, or when the antenna terminals are corroded, this results in weak and noisy signals on FM. In the discussion below, we will concentrate on antenna troubles as they appear on a TV receiver.

TV ghosts

This phenomenon is so widespread that everyone has undoubtedly seen such a picture, usually displaced to the right of the original picture, which appears as a ghost on the screen. Figure 8-10 indicates the typical origin of this type of reflection. The geometry of the transmission path also indicates the remedy. The antenna has to be

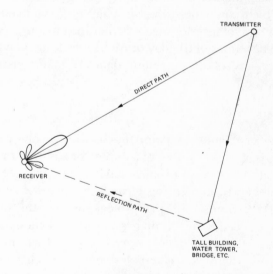

FIGURE 8-10
TV GHOSTS, CAUSED BY REFLECTIONS

rotated in such a way as to reduce the reflected signal to a minimum while maximizing the desired signal. This is not always possible, particularly in the case where different reflections from different sources reach the antenna on different channels. The only solution is to set the antenna for a compromise, to provide the best picture on those channels that the viewer most frequently watches. It is quite common in metropolitan areas to use two or three separate antennas, sometimes one outdoor and one indoor, in order to get good pictures on two or three channels. Another solution is the use of an antenna rotator, which allows you to turn the antenna by remote control to that position best suited for a particular channel.

Weak signals

We are always surprised at how often people complain of suddenly receiving weak signals on stations that were previously good, without seeing the obvious reason for it. Just a look at the antenna usually tells us that the last storm has broken off elements, ripped off the transmission line, etc. Any time you see a "snowy" picture on all channels, check the transmission line and the antenna itself.

Very often the transmission line wires are broken right at the point where the insulation is stripped back and the copper makes the connection to the terminal block, up at the antenna. This may be difficult to see from a distance and one way to check it is to disconnect the antenna leads from the receiver terminals and measure the DC resistance. Any antenna that has a folded dipole should have a very low DC resistance. With open dipole elements, unfortunately, this method cannot be used and visual inspection is required.

In homes that have CATV connections to the TV receiver, the CATV system itself should be suspected too. If other tenants report weak signals, the likely trouble is in the CATV distribution system.

Intermittents

Intermittent connections, corroded terminals, and partially cut transmission lines all cause intermittently strong and weak signals. You can prove that this defect is due to the antenna by disconnecting the transmission line from the receiver terminals and connecting a temporary, indoor-type antenna to the receiver.

The best remedy for either weak signals or intermittent antenna connections is to replace the entire installation. Whenever you replace the antenna, the transmission line should be replaced as well. The relative cost of these items justifies a single replacement job rather than two separate replacements that may only be a few months apart.

8.7 FREQUENTLY FOUND TROUBLES IN HI-FI EQUIPMENT

For troubleshooting purposes, all hi-fi and stereo equipment can be conveniently divided into the mechanical and the electronic portions. In the electronic portions we have the tuner and the amplifiers. Since most of this equipment now uses only solid-state components, the electronic troubles are much less frequent than the mechanical troubles that plague the tuning mechanism of the tuner, the electromechanical mechanism of the tape recorder, and the record player.

Mechanical trouble

Record players and tape recorders both depend on an electric motor to drive the mechanism, which rotates the turntable or moves the tape. The most frequent type of trouble is that something breaks in the drive mechanism. Record players and tape recorders both use rubber drive belts to transfer the rotation from the motor shaft to the turntable or tape transport. Because the life of these drive belts is limited, they represent a frequent source of trouble. When you can feel the vibration of the electric motor but the turntable or tape transport does not move, the drive belt is the most likely defect. You have to disassemble the machine and replace the drive belt with an exact replacement part.

The least likely part to fail is the motor itself. Very often, because of lack of lubrication or poor mechanical alignment, some of the shafts and pulleys get stuck. You will be able to find them easily when you are looking for a defective drive belt. A little fine machine oil or silicone grease usually cures that trouble.

On record players the tone arm mechanism, the cartridge, and the stylus are all subject to wear and may require replacement. On tape recorders the capstan does the driving and it may occasionally need cleaning. There is also usually a felt pressure pad that keeps the tape pressed against the playback or recording head. Check to make sure that the felt is in place and that it has not hardened or become too dirty. Remember that the magnetic head itself must be cleaned frequently. Tape recorder cleaning and lubricating kits are available at all electronic parts distributors.

Power amplifier trouble

While any of the electronics circuits in the tuner and audio equipment can become defective, the most frequent defects are found in the power amplifier and power supply portions. This is the area where components are under greatest electrical stress and where the heat generated further tends to shorten the life of semiconductors and capacitors. In vacuum-tube-type equipment all of the vacuum tubes can be expected to have a limited life and may need replacement, but in solid-state devices the power amplifier section and those power supply circuits leading to it are most likely to fail.

Most power amplifiers use push-pull circuits. If one-half of a push-pull circuit fails, the audio output will be distorted and this will be clearly noticeable to the listener. Audio transistors, particularly those used in the power amplifier section, can be tested with an ohmmeter for shorts and opens. If you have to replace a power transistor, be sure to mount it according to the manufacturer's instructions to its heat sink. When disassembling the defective unit, look for cracked or broken insulating washers and be sure to replace them as well.

Audio transformers themselves fail very rarely. Speaker voice coils can be open circuited, sometimes due to the application of sudden surge currents, but the most frequent defect in speakers is really mechanical. The voice coil mechanism can warp or twist in such a way that it rubs against its surrounding surface. Other speaker defects include loose or broken speaker cones, loose speaker mountings, or similar mechanical defects. Sometimes such defects are mistaken for power amplifier defects.

Acoustic trouble

This type of trouble is not very frequent, but we mention it here because most people overlook the possibility that an acoustic problem exists. They believe that if the audio circuits work properly, if the speakers are o.k., and if the source of the sound, tuner, record player, or tape deck produce a good sound, we should hear good sound. Unfortunately, the acoustic situation in many places is far from ideal. The same hi-fi equipment will sound differently in different acoustic

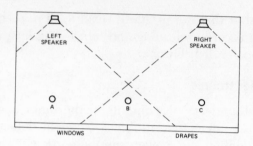

FIGURE 8-11
TYPICAL STEREO SPEAKER SETUP

environments. Drapes hung on the wall, carpets on the floor, the size of the room, the presence of doors and windows, all these will affect the quality of the actual sound that we hear.

As an illustration of an acoustical problem we refer you to Figure 8-11. The left speaker and right speaker are properly placed, so that the person sitting in the center, point B in the illustration, should hear perfect stereo reproduction. Suppose that a window covers half of the wall behind the listener and heavy drapes cover the other half. If the listener now moves to point A, he will still hear both the left and the right speaker directly, but will get strong reflections from the window surface. These reflections, depending on the dimensions of the room, may seriously interfere with the directly received sound. When the person moves to point C, where heavy drapes absorb the sound and prevent reflections, much better sound may be received.

In many large rooms there may be "dead spots," points at which, due to cancellation of reflected and direct sounds, the volume is very low. Similarly, there may be "high spots" at which particularly loud and sometimes brassy sound is heard. Be sure to take the acoustic environment into consideration when judging the performance of a hi-fi set.

8.8 A "LAST-RESORT" METHOD FOR TV AND HI-FI TROUBLESHOOTING

At the end of chapter 6 we have described a "last-resort" method for finding intermittent defects. This method applies, of course, to TV receiver and hi-fi intermittent defects as well. It happens, fortunately rarely, that troubles in TV receivers or in the electronic circuits of hi-fi systems cannot be located by any of the basic four methods. For those rare cases we suggest using the last-resort method described below. Remember, however, that this method is time-consuming, requires the correct test equipment, and cannot be performed without adequate manufacturer's technical data.

TV receivers and AM/FM tuners have a number of adjustments that control the operation of various sections, such as the tuner, the IF section, the FM detector, etc. If our four basic troubleshooting techniques have failed to locate the defect, a complete, careful and accurate alignment of the entire equipment is suggested. In the case of a TV receiver, this will require a TV sweep generator, a VTVM, and an oscilloscope. If a color TV receiver is involved, it will, in addition, require a color bar generator, which can also produce a dot and crosshatch pattern. Be sure to follow manufacturers' instructions exactly and in all respects when performing this alignment. You may find that a particular tuned circuit cannot be peaked as required by the specifications. The defect may be right there. Sometimes a broken tuning slug, a shorted tuning capacitor, or some other defect, which otherwise is very hard to find, is located by this method.

In the case of an AM/FM tuner you will need, as a minimum, a signal generator capable of covering from 400 to 1,600 KHz, from 10 to 11 MHz, and from 86 to 110 MHz. In addition, you will need a VTVM. Some TV sweep generators also cover the FM band and, in that case, an oscilloscope is also required. To check out all of the adjustments used in a stereo hi-fi audio system, an audio signal generator and a VTVM is required, although the use of an oscilloscope makes the alignment and testing a little simpler. In each case, be sure to carefully follow the manufacturer's instructions and look for a defective component whenever you cannot perform the adjustments as directed.

If the alignment of the equipment has been performed properly, you will note that there are still circuits that might contain a defect and cannot be tested by doing the alignment procedure only. To check those circuits, use the manufacturer's detailed technical data to determine the voltages at all the different test points and on all vacuum tube grids, cathodes, plates, etc., and on all transistor bases, collectors, emitters, etc. One effective method of performing this troubleshooting procedure is to measure the voltage at a particular point and then, with a colored pencil, note it down next to that point on the circuit diagram. By comparing manufacturer's data with actually measured data, you will inevitably find discrepancies that will lead to the actual defect.

9

**Troubleshooting
Digital Devices
and Computers**

9.1 WHAT YOU NEED TO KNOW ABOUT THE EQUIPMENT

Digital circuits and computers are based on a language of their own and a means of representing what goes on in the equipment that is quite different from circuit diagrams, waveforms, and voltage-resistance charts. With this language goes the knowledge of digital logic, a field that is independent of the circuits that perform these logic functions. If you have never studied these subjects, we strongly recommend two books, both published by Prentice-Hall, Inc., that will help you catch up in this area. The first book, *Digital Circuits and Devices* by T. Kohonen, gives the fundamentals that you need to know. The second book, *Fault Detection in Digital Circuits* by A. Friedman and T. Menon, shows you the specialized techniques and methods used for digital circuits.

In this chapter we assume that you have some basic knowledge of digital circuits, digital logic, and the integrated circuits that are used to perform these logic functions. To refresh your memory, however, Figures 9-1, 9-2, 9-3, and 9-4 show the building blocks of digital logic. These four circuits, shown here in the form of diode-transistor logic (DTL), perform the basic AND, OR, NAND, and NOR functions. You will understand that, regardless of the circuitry, the symbols associated with each function are what count. In an actual equipment, many of the circuits may be incorporated on a single IC and, with its internal interconnections, the IC may form a complete digital function, such as an adder, a multiplier, etc. For the troubleshooter, the important thing to remember is the truth table, which tells us what the output signal of a particular logic function should be with a given

184

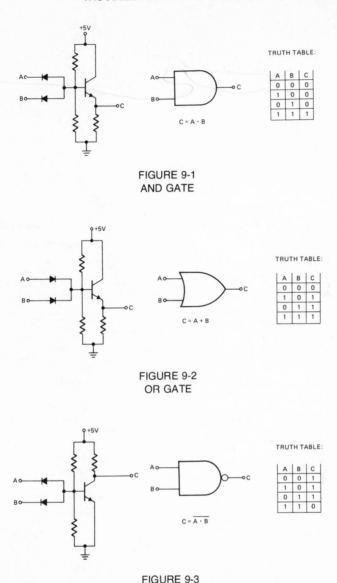

TRUTH TABLE:

A	B	C
0	0	0
1	0	0
0	1	0
1	1	1

$C = A \cdot B$

FIGURE 9-1
AND GATE

TRUTH TABLE:

A	B	C
0	0	0
1	0	1
0	1	1
1	1	1

$C = A + B$

FIGURE 9-2
OR GATE

TRUTH TABLE:

A	B	C
0	0	1
1	0	1
0	1	1
1	1	0

$C = \overline{A \cdot B}$

FIGURE 9-3
NAND GATE

input signal. There are some variations, but the basic symbols shown in this chapter are the most widely accepted ones.

In addition to the logic functions shown so far, another important element in digital circuitry is the flip-flop circuit. A variety of flip-flop

185

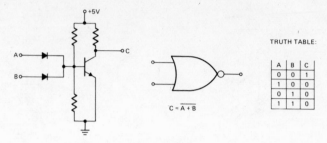

TRUTH TABLE:

A	B	C
0	0	1
1	0	0
0	1	0
1	1	0

$C = \overline{A + B}$

FIGURE 9-4
NOR GATE

circuits are in use and the basic circuits, with their truth table and functions, are shown in Figure 9-5. Flip-flops are often arranged in groups to perform such functions as shifting information from serial to parallel format, counting, storage, etc. Figure 9-6 shows the applications of flip-flops to two of the most basic combined functions as counter and shift register.

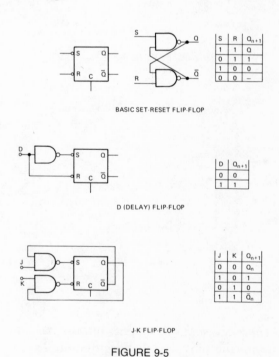

S	R	Q_{n+1}
1	1	Q
0	1	1
1	0	0
0	0	—

BASIC SET-RESET FLIP-FLOP

D	Q_{n+1}
0	0
1	1

D (DELAY) FLIP-FLOP

J	K	Q_{n+1}
0	0	Q_n
1	0	1
0	1	0
1	1	$\overline{Q}_n$

J-K FLIP-FLOP

FIGURE 9-5
TYPICAL FLIP-FLOPS

186

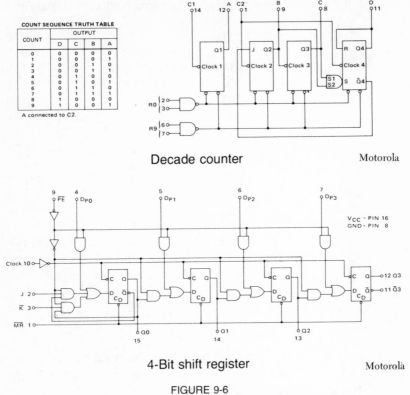

COUNT SEQUENCE TRUTH TABLE

COUNT	OUTPUT			
	D	C	B	A
0	0	0	0	0
1	0	0	0	1
2	0	0	1	0
3	0	0	1	1
4	0	1	0	0
5	0	1	0	1
6	0	1	1	0
7	0	1	1	1
8	1	0	0	0
9	1	0	0	1

A connected to C2.

Decade counter Motorola

4-Bit shift register Motorola

FIGURE 9-6
COMBINATIONS OF FLIP-FLOPS AND GATES

Digital circuits operate in the "binary" form. This refers to the fact that only two states, zero and one, can exist and this leads to the binary numbering system. To refresh your memory further, Figure 9-7 shows a brief table of binary versus decimal numbers.

In most digital devices a whole array of logic functions is used to perform the total equipment function. Very often this includes different types of storage or memory, such as a random access memory (RAM), or read-only memory (ROM), or a buffer memory. Memories can be made up of ICs, of magnetic core matrices, of rotating magnetic discs, drums, and even magnetic tape. These different types of memories have different characteristics as concerns their access time, the method of entering information or reading information out, whether the memory is destroyed when power is removed from the

Decimal number	Binary number	Decimal number	Binary number
0	0	17	10001
1	1	18	10010
2	10	19	10011
3	11	20	10100
4	100	21	10101
5	101	22	10110
6	110	23	10111
7	111	24	11000
8	1000	25	11001
9	1001	26	11010
10	1010	27	11011
11	1011	28	11100
12	1100	29	11101
13	1101	30	11110
14	1110	31	11111
15	1111	32	100000
16	10000		

FIGURE 9-7
DECIMAL AND BINARY NUMBERS

equipment, etc. Books have been written about each of the different types of memories and if you want to delve deeper, the two books we have listed above are a good start.

Now that you know what fundamental knowledge you have to bring to any troubleshooting efforts on digital equipment, let us review the type of material you can expect to find in the manufacturer's trouble-shooting data. The most important and informative piece of information is the block diagram and the logic diagram. Figure 9-8 shows the block diagram of a relatively simple device, a digital clock, such as is used in providing time-of-day and the date on a computer system or in a TV broadcast station. This block diagram tells us, at least, what the major functions are. We can see that, in this particular case, a defect that affects only the hour indication will inevitably also affect the calendar portion, though not necessarily the minute and seconds portion of the digital clock.

Figure 9-9 shows the logic diagram of a small portion of the block diagram of Figure 9-8. It covers the nixie tubes, the drivers, and counters for the second and minute portion of the calendar clock. We shall see later how this information allows us to perform the digital version of the signal-tracing method.

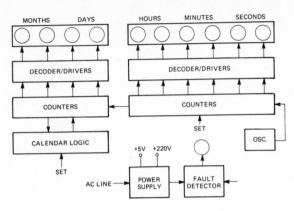

FIGURE 9-8
BLOCK DIAGRAM OF DIGITAL CALENDAR CLOCK

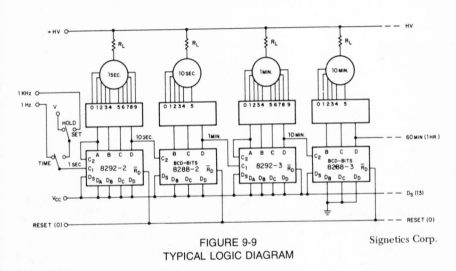

FIGURE 9-9
TYPICAL LOGIC DIAGRAM

Signetics Corp.

In an equipment like the digital clock we have used as an example where continuous digital signals are applied, it is important to know the timing relations between these signals. This information is usually provided by means of a timing diagram, like the one shown in Figure 9-10. If we use the signal-tracing technique here, it would be helpful to have an oscilloscope, which can display several of these waveforms at the same time. More about this later.

Two other troubleshooting aids are frequently available with digital equipment that performs computerlike functions. One is the flow chart, an example of which is shown in Figure 9-11, and the other is a

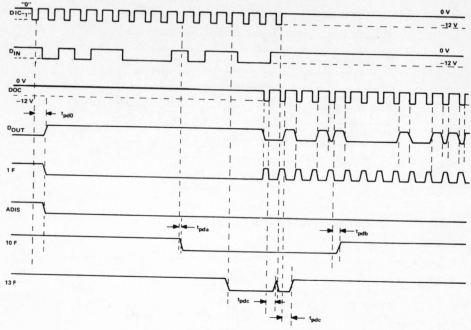

FIGURE 9-10

TIMING DIAGRAM, NEGATIVE LOGIC, MOS ICs

Texas Instruments, Inc.

diagnostic program. Flow charts tell us the sequence of events that must occur in normal operation. Diagnostic programs usually require a knowledge of programming and are not suitable for ordinary troubleshooting. Sophisticated digital equipment, such as mini-computers or computers, have a means of entering data and programs, such as a teletypewriter or a card reader. The manufacturer of the machine usually supplies a diagnostic program, either in the form of punched paper tape, IBM cards, or magnetic tape, and by entering this program into the machine the resulting output, in the form of a printout or similar indication, will tell us in what portion of the machine the defect is located. The actual troubleshooting then focuses at that point.

If we have convinced you that troubleshooting digital equipment is an extremely difficult task, one that the average technician cannot hope to conquer, we have overstated our case. In actual practice, trouble-shooting digital equipment is quite simple because almost all digital equipment is built in the form of plug-in PC cards with ICs mounted on them. The PC card shown in Figure 9-12 is typical. If you refer to

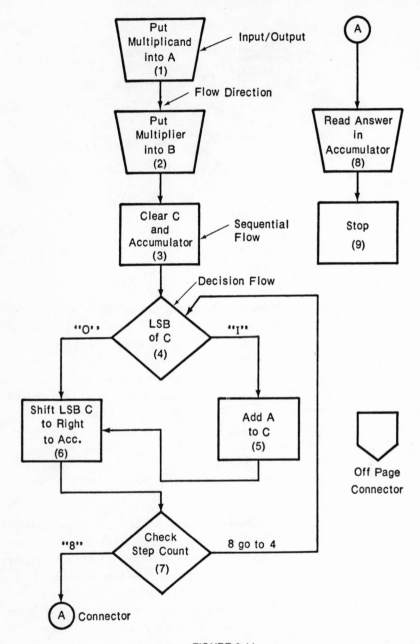

FIGURE 9-11
TYPICAL FLOW CHART

191

FIGURE 9-12   Chrono-Log Corp.

the manufacturer's instructions, taking as example the digital clock, you will find clear instructions as to which PC card to exchange if either the seconds, minutes, hours, etc., are wrong. Just as in vacuum tube equipment, ninety percent of the troubles could be cured by exchanging tubes, so, in digital equipment, plug-in PC cards account for a large proportion of the defects. For this reason, manufacturers of this equipment frequently supply plug-in cards as spares and very often a single type of plug-in card is used many times in the equipment. To further help us troubleshoot their equipment, most manufacturers also provide test points and, in many cases, visual indicators, such as light-emitting diodes (LED), to show when trouble occurs.

When a defect has been isolated to a plug-in card, the usual procedure is to replace this card with one known to be good. The defective card can either be returned to the manufacturer for repair, or, in some instances, further troubleshooting down to the defective IC is possible. As we show you in the rest of this chapter, if you know the fundamentals of digital logic and if you have mastered the four basic methods of troubleshooting, you can, with confidence, set out to troubleshoot digital devices and even computers.

9.2 PECULIARITIES OF DIGITAL CIRCUITS

The uniqueness of digital circuits lies in the fact that a particular stage is either "on" or "off." When we look at this statement a little closer, the question arises as to just what is "on"? Suppose we are dealing with a transistor-transistor logic (TTL), which is one of the most widely used logic families. Inevitably, the supply voltage to these ICs is +5 volts. If positive logic is used, this means that the "one" is a level of +5 volts and the "zero" is ground. The manufacturer's data for this family of ICs gives a specific value of 3.35 volts as the "on" and +0.35 volts as the "off" tolerance limits. If we add to these values the expected tolerance of a typical volt-ohmmeter, this means that, for practical reasons, we can assume that any time the voltage is 3.5 volts or higher we have a "one," and if the voltage is 0.5 volts or less we have a "zero." For different logic families, such as MOS, these values are different. In any event, we can be sure that something is very wrong if we measure a voltage that is between the values of zero and one. Any time you measure two volts on a TTL logic chip, you know that you have located a defect.

One of the requirements for digital circuits that becomes apparent from the above discussion is that so-called "hot grounds" can cause serious troubles in digital circuits. Power supply transients can also cause false logic actions.

As we shall see in more detail in paragraph 9.5, checking the logic levels can only be done on static testing. More digital circuits are controlled by rapidly occurring pulses and, in these, different criteria become important. Figure 9-13 shows a typical pulse, such as would be used to actuate a counter or some other logic element, and the parameters that are important in digital circuits. The height of the pulse corresponds to the logic voltage level. The rise time determines

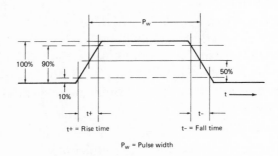

FIGURE 9-13
PULSE TIME PARAMETERS

the speed with which this voltage level is reached. The fall time describes the speed with which the voltage level is reduced to ground.

Different types of logic families are designed to operate at different speeds. Many of the high-speed logic families, such as emitter-coupled logic (ECL), will not operate properly when the rise and fall times of the control pulses are too large. Slower speed logic, such as resistor-transistor logic (RTL), cannot operate when the rise time and fall time are too short. We often speak of 1 MHz, 10 MHz, or 50 MHz logic. By this we mean the highest pulse rate with which this logic can operate. A 1 MHz pulse rate implies that, in the case of a square wave, each half takes approximately 0.5 microseconds. This further means that the rise time should be no slower than about 0.1 microsecond. A microsecond is one millionth of a second and is frequently denoted by 10^{-6} seconds. One one-thousandth of a microsecond, formerly called a millimicrosecond, is now generally called nanosecond; 0.1 microseconds therefore equals 100 nanoseconds. Rise and fall times are important characteristics that will be found in the manufacturer's data for all ICs and can be used either in analyzing a particular defect or in determining a suitable replacement IC.

We know what happens when the proper pulses arrive at two inputs to a particular logic gate. Figure 9-14 shows two inputs to a simple AND gate. If signal B is delayed from signal A as shown, the output of the AND gate will be A • B pulse. Suppose, for some reason, one pulse occurs after the first pulse has already disappeared. This may be due to some extra time delay elsewhere in the system. This is called a race problem because we can conceive that one pulse train has outraced the other. The result is shown as A • C. Every stage of digital logic delays the pulse train a certain amount, because it takes a finite, however small, amount of time to turn the transistor on and to turn it off again. This so-called "stage delay" is usually measured in nanoseconds and is normally accounted for in the overall logic design. Leaky capacitors or other defects can change the stage delay somewhat in the system and cause race problems, which are extremely difficult to find.

Digital logic circuits often are connected to a number of other digital logic circuits, forming a parallel load to the output of one stage. The IC manufacturer's data indicates how many loads of the same logic family can be connected to a particular output. This is called "fan-out"

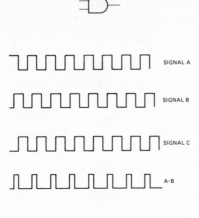

FIGURE 9-14
EFFECT OF PULSE DELAY

and varies among different circuit elements. A typical flip-flop stage, for example, may have a fan-out of two. This means that two other flip-flops or gates of the same logic family can be connected to this flip-flop without loading it down. If four or five such circuits are connected, the flip-flop may be loaded too much and may not operate under all conditions. In most ICs the outputs are special driver stages that are designed to drive a larger number of circuits. Typical fan-outs for driver stages are in the order of ten or twelve. A less frequently used parameter is the "fan-in," which determines the number of inputs that can be connected to a particular terminal. For applications where fan-ins or fan-outs must be greater, there are special "expander" ICs that are, in effect, driver circuits and can handle up to a dozen or more inputs or outputs.

9.3   HOW TO USE THE SYMPTOM-FUNCTION TECHNIQUE

As we have learned in the case of TV and hi-fi equipment, the symptom-function technique usually allows us to isolate a defect to a particular section. This also holds true for most, though not all, digital equipment. If we refer to the block diagram of the digital clock shown in Figure 9-8, we can clearly see that defects occurring in the hour counter will have an effect on a calendar portion but should not affect the minute or second counter. If the minute and second displays as well as the calendar displays are correct and only the hour display does not work, the trouble is most likely in the particular display or its driving circuits. This type of symptom-function troubleshooting can easily be done on most equipment.

If we are trying to troubleshoot one of the new digital calculators, where all of the circuitry is contained on one or two LSI ICs, the symptom-function technique may be of little help. We can, by static tests, check that each of the displays works correctly, we can take ohmmeter and voltmeter readings on the push buttons that control the ICs, but we can only replace the ICs themselves if the calculator consistently produces the wrong results.

In digital computers, whether micro-, mini-, or full-sized computer, only a thorough and detailed understanding of the entire computer architecture would enable a person to use the symptom-function technique. This type of knowledge is usually reserved for the computer designer, but the manufacturer provides some kind of step-by-step troubleshooting procedure in which symptoms and functions are grouped to lead to the most likely defects. Figure 9-15 shows a portion of this procedure provided by the manufacturer for the digital clocks described earlier in this chapter. This is another reason why we strongly advise you not to attempt troubleshooting any digital equipment without having the manufacturer's manual on hand. In many instances the step-by-step procedure also directs us to replace certain plug-in PC modules in case of some specific symptoms. In digital circuits this is usually the extent to which we can use the symptom-function technique.

9.4 HOW TO USE SIGNAL TRACING

In digital equipment the signal-tracing method can be applied in two different ways. Until we narrow the search down to a few specific components, the first method, dynamic signal tracing, is preferred. The second method is static signal tracing, checking DC logic functions.

As a brief example, let us assume that the symptom-function technique has isolated the fault to the minutes and seconds portion of the digital clock. Figure 9-8 shows the block diagram and the detailed logic circuit is shown in Figure 9-9. To begin with, we need a good oscilloscope. At the left in Figure 9-9, we see the 1 KHz and the 1 Hz signal. The latter signal drives the second counter, which, in turn,

STEP	Perform the Following	Observation
26	Place oscilloscope ground probe on J1, pin 1. Place oscilloscope test probe on serial output pin of bit to be tested. For example, J1 pin 71 is the serial output for the "1" bit. Ground the strobe input for all digits *not* under test. For example, to test Units Seconds, ground the following strobe inputs on J1:	

pins 38, 46, 54, 62, 70, 69,
67 and 65

Observe the value of the output for the bit under test while manually stepping the digit under test through its complete count cycle. The following table lists all values:

Count of Digit Under Test	Corresponding BCD code	Output Pin 75	Voltage on J1, Pin 73	Pin 77	Pin 71
1	0001	5	5	5	0
2	0010	5	5	0	5
3	0011	5	5	0	0
4	0100	5	0	5	5
5	0101	5	0	5	0
6	0110	5	0	0	5
7	0111	5	0	0	0
8	1000	0	5	5	5
9	1001	0	5	5	0
0	0000	5	5	5	5

STEP	Perform the Following	Observation
27	Repeat step 26 for each digit, and each serial output line.	
28	*Pulse Rate Outputs* Connect the oscilloscope probe to the following pins on J1 and with the clock running, observe the pulse rate outputs:	The pulse rate outputs have a 20% on (ground level) and 80% off (+5 V level) duty cycle.

pin 59   1 KHz
pin 57   100 Hz
pin 53   10 Hz

STEP	Perform the Following	Observation
29	*Power Failure Output* Connect the oscilloscope probe to the Power Failure Error bit output, J1, pin 55. Turn off AC power to clock, then turn on AC power.	The Power Failure Error bit output should be at the +5 V. level signifying power failure. Front panel error lamp should be on.

FIGURE 9-15

drives the binary-to-decimal converter, which illuminates the nixie display tube. First we use the oscilloscope to check for the presence of the 1 Hz pulse signal. If the 1 Hz signal is not available or is distorted, the defect may be found in previous counter stages. For test and clock setting purposes, the switch permits the 1 KHz signal to be connected to the counters. This means that the 10-second lead, going from the unit counter to the tens counter, will instead have 100 Hz on it. The tens counting stage is connected to divide by 5, so that the signal emerging at point D of the second counter will be a 20 Hz pulse train. These individual pulse trains and their levels at points A, B, C, and D respectively at each counter are checked with the oscilloscope and compared with the information contained in the step-by-step procedure shown in Figure 9-15.

In more complicated digital equipment, dynamic signal tracing may require that the input signals be provided by special "word" generators, while separate clock generators simulate the normal sequence of signals. Word generators provide pulse trains with a selectable sequence of "ls" and "0s." One type of oscilloscope used in digital troubleshooting has the capability of storing one such pulse train on the CRT. We can connect such a "storage" oscilloscope so that a test "word" is entered at one end of the machine and the screen then shows the resultant pulse train. By analyzing what has happened to the test word in terms of binary math, a skilled digital troubleshooter can often isolate the point where the defect occurred.

A much simpler method of signal tracing digital equipment makes use of static or DC operation. In effect, we test each individual gate, flip-flop, etc., by this method and we need only a few suitable clip leads and some means to indicate whether a "1" or a "0" appears at the output. Figure 9-16 shows a simple logic probe circuit, which can be connected to the +5 volt power supply of the equipment under test and which can be moved from point to point. When connected to a "1," the light-emitting diode (LED) will light up. You can test a particular logic gate simply by connecting its inputs, by means of your clip leads, so that a "1" should appear. Referring to the AND gate in Figure 9-1, for example, you could simply tie both A and B terminals to the B+ and make sure that a "1" appears at terminal C. To fully check the operation of this gate, the A and B terminals can be connected as indicated by the truth table.

This same method can be expanded by using the circuit of the logic

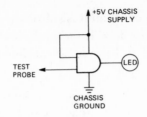

FIGURE 9-16
SIMPLE LOGIC PROBE CIRCUIT

test probe in Figure 9-16 in a multiple arrangement. In computer labs it is common practice to have an array of 10 or 20 LEDs or other indicators connected to driver circuits, which, in turn, are hooked up to a shift register, counter, etc., which is part of the electronic equipment under test. In this manner, the binary state of the shift register under test is always available and, by varying the input to the circuit, such as setting up the word generator differently, we can obtain clues as to the location of the defective stage.

9.5 HOW TO FIND THE DEFECTIVE COMPONENT

The voltage-resistance measurement technique is of limited use in digital equipment, except for failures occurring in the power supply. In the digital circuitry, most of the resistance measurements are meaningless when power is turned off because the ICs, which form 90 percent of all components, are effectively open circuits. Voltage measurements are equally meaningless, because the logic power supply is usually well regulated and, even if one IC draws excessive current, the logic voltage may not show any change. Complete short circuits usually caused the fuse or circuit breaker to disconnect the power supply voltage. Even the ohmmeter test for a short circuit on a particular PC board is not very useful because it is usually difficult to isolate the short circuit to the component causing it. To do this would require cutting some of the conductors that make up the printed circuit on the PC board.

Chapter 4 contains a discussion on the limitation of the voltage-resistance technique as well as the substitution technique. We suggest that you look over chapter 4 again to see how ICs can be tested, how they are likely to fail and how you can use IC data to analyze possible failures. One failure mode that is relatively frequent is that the IC draws more current than it is supposed to. This can be tested if you can unsolder the E_{CC} lead from the IC and connect a milliameter in series with that lead and the logic supply voltage.

In some ICs a measure of testing is possible, such as in the case of the four-bit shift register or the decade counter shown in Figure 9-6. If you disconnect the clock and instead operate either of the circuits on a DC basis, you can measure the voltage outputs at each of the four flip-flops. If the voltage falls beyond the tolerance levels for "1" and "0," then you know that you have found a defect. Unfortunately, regardless of which stage the defect is in, the entire IC must be replaced.

In digital equipment capacitors are mostly used to isolate B+ and certain signal sources. These capacitors then act mostly as bypass and filter capacitors. To make sure that a capacitor is not open, you can shunt it temporarily with one known to be good. If it is shorted, removing one of the two leads and substituting a new one is the accepted method of testing for short circuits.

In the vast majority of digital equipments ICs are soldered to the PC board. To remove an IC requires either the availability of the proper desoldering tool, or you have to use the time-consuming "pin-by-pin" method. You have to clip off all of the pins of the IC and remove the body. Then you can withdraw each pin individually while heating the land to which it is soldered. After all pins are removed, be sure to clean each land carefully, cleaning away excess solder and making sure that each hole is clear. Inspect the space between lands to be absolutely sure that no solder bridging exists. Now you can insert the replacement IC, being careful that none of the pins are bent or broken. Each pin has to be resoldered separately and you must make sure that no bridging exists after the job is completed.

9.6 FREQUENTLY FOUND TROUBLES IN DIGITAL EQUIPMENT

After reading of the difficulties usually encountered in replacing ICs, you will be glad to learn that ICs are not among the most frequent failures in digital equipment. As a matter of fact, ICs are usually last on the list.

The most frequent failures are due to poor connections in one of the many different connectors, cables, and other wiring. Large digital equipments have an astonishing number of interconnecting wires. If a particular equipment consists of twenty or thirty PC cards, each card is likely to fit into a connector having between 30 and 80 pins, and wires have to go from one connector to the other. In some equipments the interconnecting wiring itself is provided by means of a so-called "mother" board, a printed circuit that consists of only interconnections and does not contain any components. While a conductor on a PC board or a point-to-point wire is generally a very reliable part, the point at which connectors meet is a frequent source of trouble. Wherever we have a wire soldered, wire wrapped, swaged or crimped to a terminal, the likelihood of a defect is always much greater than in the wire or printed conductor itself.

The most frequent offenders are the PC board connectors themselves. Many equipments, like the digital clock used as earlier example, use so-called edge connectors. As seen in the top of Figure 9-12, the printed conductors are arranged in a pattern on the edge of the PC board. Spring-loaded contacts in the connector then engage these printed conductors. PC connectors may look good but sometimes, because of wear and tear on the PC card itself, the mating connection is not reliable. One easy way of testing this is to move and wiggle the PC card and see if the defect still exists, if it gets worse, or if the wiggle makes no difference whatever. The printed contacts on a PC card can easily be cleaned with cleaning fluid or a commercial contact cleaner. Under no circumstances should contacts be scraped or sanded if they appear corroded. Do not use silverpolish or any chemicals for this purpose.

To make sure that a particular PC connector is good, look into the connector, with the PC card removed, to see if there are any small pieces of metal, solder drops, etc., that may be lodged in the plastic connector, either shorting out to adjacent pins or else blocking the spring action on a particular contact. Visual inspection also can show where a spring contact is worn or broken off. In some types of connectors it is possible to replace individual contacts, but in most cases the entire connector has to be replaced.

Next to the PC board connectors the most frequent trouble spots in digital equipments are the cable connectors, particularly those in which the wire has to be soldered into each connector pin. Wires have a tendency to break off at the point where the solder from the

201

pin stops. This is usually also the point where the insulation stops, and breaks at that point are often very difficult to see. The best way to locate such a defect is to check out a complete cable, wire by wire, with the ohmmeter and jiggle each lead while testing for continuity.

The third most likely source of trouble is the power supply itself. Power supplies used in digital equipments are usually well regulated, filtered, and protected against short circuits and against overvoltage or AC line transients. A component failure in any of these circuits, however rare or unlikely, can have disastrous consequences. Earlier in this chapter we have discussed the results of ripple voltage riding on the logic supply voltage and thereby changing the logic level tolerance. An open filter capacitor, a defect that is not at all infrequent, can cause this defect. Power supply regulation depends usually on the use of pass-transistors, which pass all of the power supply current. The heat generated in these transistors tends to shorten their life, even though the proper heat sinks are used, and failure of the pass transistor is another relatively common defect. When the pass transistor opens there will be no power supply voltage output. If the pass transistor shorts, which is relatively rare, the voltage will be unregulated and this means that when a large number of logic elements change state simultaneously, a power supply transient will be produced, which can cause logic errors.

The short circuit protection components fail relatively rarely. If they fail in such a manner as to cause the circuit breaker or fuse to remove the power supply voltage from the equipment, repeated setting or replacement of the fuse will quickly indicate this type of failure. If the short circuit protection components fail in such a way that they no longer protect the power supply against the short circuit, we will only become aware of this after a short circuit has occurred and some of the power supply components have been damaged. A much more insidious type of defect is failure of the overvoltage protection circuit. When this portion of the power supply fails, transients due to the AC power line can briefly exceed the regulated power supply voltage and this can burn out a large number of ICs. Fortunately, this type of defect is extremely rare. If, however, your troubleshooting shows that a number of ICs are defective, this is an indication that you should check the overvoltage protection circuits.

As was described in more detail in chapter 4, failure in the connections of the IC package itself are not very frequent but they can be very hard to find. This is particularly true in the case of intermittent failures as discussed in more detail in chapter 6. As far as the trouble-shooter is concerned, it does not matter whether the IC itself or only the external pin connections are defective. Replacement of the entire package is required.

The original equipment designers usually make sure that proper electrical grounding is provided in digital equipment. Very often a ground path, in the form of a copper strip, runs alongside a row of connectors and short wires are then brought to each PC board connector. On the PC board itself it is common practice to use at least two or more connector pins as grounds. When troubleshooting digital equipment, particularly when so-called extender cards are used, ground connections are often neglected and subsequent problems are compounded by the poor grounding practice. While grounds themselves are not a frequent source of trouble in digital equipments, be sure that the proper ground connections exist in the setup that you use for trouble-shooting.

9.7 A LAST-RESORT METHOD FOR DIGITAL TROUBLESHOOTING

As in the case described for TV receivers and hi-fi equipment in the previous chapter, the last-resort method suggested here is time-consuming and requires special test equipment in most cases. Before using this method, we suggest that you go over everything that you have done so far and write down just what tests you have made and what the results have been. Review your notes and, hopefully, you will see where you have made a mistake, what you have overlooked, or what clues remain to be explored. The joke "as a last resort read the instruction book" is only too true, particularly for digital equipment. Check over the manufacturer's instruction book, any trouble-shooting guides, or other information that may be available. It would be foolish to waste the time required to go through the last-resort method if you have missed locating the trouble by some minor error.

The manufacturer's instructions usually contain some information as to the test signals to be used at different points and the expected results. In the case of mini-computers this usually means inputs from a card reader, papertape reader, etc. Since we cannot be sure that the electromechanical device operates correctly, we may have to set up a word generator with a code equivalent to the data from the punched card, papertape, etc. Once we have set up the proper input signals in the form of the required digital pulse trains or input register settings, we now need to record the waveforms at all test points. One way to accomplish this is to connect the oscilloscope, in sequence, to each of the clocks in the equipment. If a single master clock is used from which all subsequent clocks are counted down, a multitrace oscilloscope is helpful. Any linear graph paper, such as the kind that has ten squares to the inch, for example, is handy to record pulse waveforms. It is important that a single timing reference, such as the start of the oscilloscope sweep, be used. After you have recorded all of the clocks in the equipment, you can compose, by reference to the functional block diagram, a timing diagram of the entire equipment. You can now compare the observed timing diagram with the one presented in the manufacturer's information.

Similarly, you should record the status of shift registers, counters, adders, etc., and similar functional logic blocks with the known test word input. Again, this information can be compared with the manufacturer's data.

Let us return to the simple digital clock used as an example earlier in this chapter. If you refer to the block diagram in Figure 9-8, it will become apparent that if you set the counters of the seconds, minutes, and hours portion to read all "1" (11:11:11), then you can see whether the output of each individual decoder and driver is indeed indentical. Referring to the logic diagram in Figure 9-9, you would set the switch to "hold" while you check the outputs of the individual counters and the nixie drivers. With the switch turned to "set," you can enter a single pulse and check to see whether this sets only the 1 second counter from 1 to 2. Clearly it is tedious to go through all the steps of this type of logic testing, but remember, it is a last-resort method to be used if all other methods fail.

In more complex digital equipments you will be able to isolate the trouble to one particular section. The last-resort method is then applied only to the section to which the trouble has been isolated. Eventually, with a lot of hard work and attention to detail, you will find the discrepancy between the data observed and the data given by the manufacturer. In very tough cases, once you have isolated the defect to a number of ICs, it may be simpler and faster to replace them all than to spend the additional last-resort time to track down the one individual defect.

10

Troubleshooting Industrial Controls and Instruments

10.1 WHAT YOU NEED TO KNOW ABOUT THE EQUIPMENT

As in the case of TV, hi-fi, and digital equipment, you should understand the fundamentals and some of the special aspects of industrial controls and instruments if you want to troubleshoot them. All industrial controls and instruments have certain basic characteristics in common. As shown in the block diagram in Figure 10-1, they all consist of an input device, a sensor or transducer, a control or functional portion, and an output device or actuator. The input device senses some physical characteristic, such as motion, temperature, light, humidity, air pressure, liquid flow, chemical change, etc. These physical quantities are always translated into an electrical analog and the device that does this translation is called a transducer. The thermostat that controls the temperature in our houses is a transducer, and so is the tachometer, which indicates the speed or rotation, the photocell, which changes variations of light into variations of electric current, and many more.

The control portion of the system acts on the input signal in order to produce a control output. This output can be a meter reading or it can be in the form of a physical action. In the case of our thermostatically controlled heating system, this physical action means turning on the heat. In the case of a photoelectric cell, which senses the loss of daylight, the control voltage is used to actuate a relay that turns on the light in the house. In each case we can see that some kind of actuator or output element is required. Because a control function is performed electrically, this output element must be able to change electric energy into some of the physical parameters mentioned be-

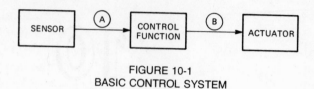

FIGURE 10-1
BASIC CONTROL SYSTEM

fore. Typically, a solenoid or motor is used, which changes electric current into a magnetic field, which is then changed into mechanical motion.

Any system in which input and output are related by means of a control function can be considered a servo system. Two basic servo systems are possible—the open loop and the closed-loop servo system. An example of an open-loop system would be a timer-controlled traffic light. Here the output of the system, the traffic light, is independent of the element it tends to control, such as the actual traffic. The output is only determined by the timing. A closed-loop servo system always requires a connection between input and output, generally referred to as feedback. A typical example of a simple servo system is shown in Figure 10-2, which illustrates a constant speed drive for some kind of a drum. The speed control potentiometer is used to set the reference level of the differential amplifier that drives the motor. In this example, the transducer is the tachometer, which generates a voltage depending on the speed of the drum. As long as the tachometer output and the reference voltage are the same, a constant voltage will be supplied to the motor. If the drum slows down for any reason whatever, the reduction in tachometer output

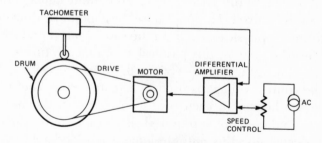

FIGURE 10-2
CLOSED-LOOP CONTROL SYSTEM

will cause the differential amplifier to put more current into the motor, which, in turn, will tend to speed it up until the drum speed is again at the desired level. The feedback element here can be considered the tachometer, while the motor is clearly the actuator. The speed-control potentiometer and differential amplifier are the control elements in this servo loop.

To help those of our readers who may have forgotten the principles of electric motors, we present Figure 10-3, which contains a table of motor types and the basic circuits of each. Whenever a defective motor is suspected, you know that ohmmeter checks give only an approximate indication of whether windings are open or shorted. Partially shorted windings in a motor, something that occurs fairly often, cannot be checked by ohmmeter tests alone. Remember that the electrical portion of the motor can be damaged if there is something wrong in the mechanical portion. If the drive mechanism is stuck, for example, or if the motor shaft has been bent, the windings can easily burn out. In any event, if a motor is defective, it usually has to be replaced.

Next to motors, one of the most widely used electromechanical devices in industrial controls is the relay. Relays come in such a wide variety of configurations, sizes, and contact ratings that no one single manufacturer produces all of them. In checking relays we only need an ohmmeter to determine whether the solenoid coil is open or shorted and whether the contacts make or break as they should. As a handy reference, Figure 10-4 contains the most commonly used relay contact arrangements and their special nomenclature. Some relays may have only "normally open" contacts, others may have a mixture. Some relays operate on AC, some on DC. Some of their contacts are of the "make-before-break" type and others use the reverse sequence. In replacing a defective relay, it is absolutely essential to use an exact replacement type. Relay contacts themselves can often be repaired, at least temporarily. When the contacts are slightly corroded, they can be cleaned with the proper tools. Sometimes we can bend the arms of relay contacts to assure proper operation but this is not usually recommended. Because relays are relatively inexpensive, a replacement is the preferred method of repairing the trouble.

The table in Figure 10-5 shows a list of the different types of transducers, actuators, and controls that you are likely to find in industrial control and instrumentation equipment. Space does not permit us to

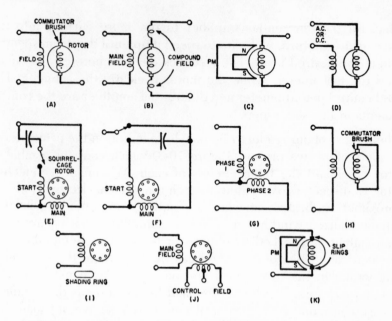

TYPE & FIG. REFERENCE	STARTING TORQUE	SPEED CONTROL	OPERATING FEATURES	TYPICAL APPLICATIONS
D.c. shunt (A)	Medium	Thyratron or voltage control	Adjustable speed; constant torque or constant power	Pumps, conveyors, wire and paper winding
D.c. compound (B)	High	Usually not used	Speed adjustable over small range; high but varying torque	Flywheel drive, shears, punch presses, hoists
D.c.-PM field (C)	Low	Power tubes or transistors		Fans, blowers, battery-operated devices
Universal series d.c. or a.c. (D)	Very high	Thyratron, saturable reactor, series re-sistor	High speed; high efficiency	Hoists, cranes, vehicles, hand tools, appliances, general utility
Capacitor start a.c. (E)	Very high	Saturable reactor	Limited range of speed control as torque drops with voltage	Compressors, pumps, blowers
Capacitor running (reversible) (F)	Low	Usually not used	Speed varies greatly with load	Fans, blowers, centrifugal pumps
Squirrel-cage induction (poly-phase) (G)	Depends on type used	Saturable reactor, resistors	Available in six classes of performance characteristics	General-purpose industrial motor used as main power source for heavy machinery
Repulsion-start, induction-run (H)	Very high	Usually not used	High starting-current surge	Pumps, compressors, conveyors
Shaded pole (I)	Very low	Usually not used	Relatively inefficient, but low in cost	Fans, blowers, heaters, phonographs
Servo (J)	High	Power amplifier, saturable reactor	Accurate control through special control winding	Positioning systems, computers
Synchronous (K)	Low	None	Constant speed depends on number of poles and line frequency	Clocks, timers, blowers, fans, compressors

Illustration courtesy of *Electronics World*

FIGURE 10-3
TABLE OF MOTOR TYPES

From *Buchsbaum's Complete Handbook of Practical Electronic Reference Data*, Walter H. Buchsbaum, ©1973, Prentice-Hall, Inc., p. 425.

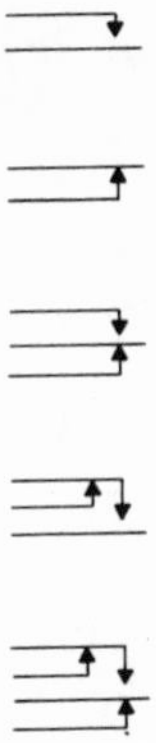

	FORM "A"	S.P.S.T., NORMALLY OPEN
	FORM "B"	S.P.S.T., NORMALLY CLOSED
	FORM "C"	S.P.D.T.
	FORM "D"	MAKE BEFORE BREAK
	FORM "E"	BREAK, MAKE BEFORE BREAK

FIGURE 10-4
STANDARD RELAY CONTACT ARRANGEMENTS

TRANSDUCERS

Mechanical:
    Limit switch
    Variable resistance
    Capacitive
    Inductive
    Differential transformer

Temperature:
    Bimetallic
    Thermistor
    Solid state

Photoelectric:
    Photoresistive
    Photovoltaic

Humidity:
    Hair, gut, films

Air pressure:
    Diaphragm
    Burdot tube

Liquid flow:
    Turbine
    Venturi
    Ultrasonic

Chemical:
    Ion exchange
    Battery action

ACTUATORS

Solenoids
Motors
Pneumatic actuators
Hydraulic actuators

CONTROLS

Relays
Magnetic amplifiers
Power amplifiers
Function generators
Fuses
Circuit breakers

FIGURE 10-5
ELEMENTS OF INDUSTRIAL CONTROL SYSTEMS

discuss each of these types in detail but you can learn more about transducers, actuators, and controls by referring to the following books:

Handbook of Transducers for Electronic Measuring Systems, H. Norton, Prentice-Hall, Inc., 1969.

Industrial Control Electronics, M. Mandl, Prentice-Hall, Inc., 1961.

Electric Machinery and Control, I. Kosow, Prentice-Hall, Inc., 1964.

Electromechanical Devices for Energy Conversion and Control Systems, V. Del Toro, Prentice-Hall, Inc., 1968.

10.2 CHECKING INPUT AND OUTPUT SIGNALS

If you have an instruction book handy, the best way to start any troubleshooting job is to skim through the instruction manual. Industrial control and instrumentation equipment is no exception to this rule. The manufacturer's instruction manual will tell you, at least roughly, what the particular equipment does, how it does it, and what the basic elements are. As soon as you know what type of sensor or transducer is being used, you can get some idea, even without the instruction book, as to what type of signal to expect. Similarly, if you know what the actuator is and how much force the actuator has to provide, you can get a pretty good estimate of the output signal.

The block diagram of the basic control system shown in Figure 10-1 illustrates the two most important points for the troubleshooter. Point A is the input to the control function and point B the output of the control function. By checking these two points you will be able to determine relatively quickly what is basically wrong with the industrial control and instrumentation equipment that you are working on. By checking at point A, you can tell whether the sensor output signal is correct. You can then check at point B to determine if the control function performs its part correctly. You can also tell, by checking at point B, whether the actuator receives the proper electrical input. You can then observe, either visually or by checking the output of the actuator in some way, whether it functions correctly.

Look at the table in Figure 10-5. Among the mechanical transducers only the limit switch and the variable resistance transducer can work properly with DC. All other mechanical transducers, the capacitor, inductor, and differential transformer, require AC. As a matter of fact, AC voltages are impressed on almost all types of industrial sensors because it is so much easier to amplify small differences in AC voltage. DC amplifiers are inherently unstable. Where DC voltages must be used, such as in chemical transducers, they are frequently converted into AC by means of so-called chopper circuits.

Going down the list in the table in Figure 10-5, we can see that the bimetallic temperature sensor, the thermostat, the thermistor, and solid-state devices can all operate on either DC or AC. Photoresistors usually operate on AC while the photovoltaic cells generate a voltage corresponding to the illumination. Humidity-sensing devices can operate either on AC or on DC, especially since some of them translate changes in humidity into mechanical motion, which is then changed into resistance. The same thing applies to air pressure and many liquid-flow transducers. Motion of the diaphragm or of the Burdot tube is usually translated into a change of resistance, either by mechanical linkage to a potentiometer or by means of a strain-gauge transducer. Turbine and venturi-type flow meters generally operate on AC and the ultrasonic flow meter, of course, always operates on varying frequency AC. Most chemical transducers, such as the pH meter, operate on the basis of ion exchange and this invariably means a small direct current.

In general, the best way to check an AC input signal is with the oscilloscope because it will show us the frequency of the signal, its amplitude, and any possible distortion. The impedance levels of some transducers can be fairly high, and it is therefore important that the oscilloscope probe contain isolating resistance to avoid loading down the transducer itself.

DC measurements of the transducer output always require a high impedance probe and a very sensitive meter. The volt-ohmmeters with which we ordinarily measure DC are not capable of measuring millivolts or even microvolts. If you expect to troubleshoot industrial control equipment and instruments that use DC transducers, you may have to invest in a "millivoltmeter," a piece of equipment not ordinarily found on the troubleshooter's bench.

Measuring output signals is usually much less of a problem. Solenoids and motors must have the rated voltages applied to them in order to operate. These voltages may range from five to several hundred volts, all within easy range of a standard volt-ohmmeter. To get an idea as to what voltages, AC or DC, are applied to which portion of the motor, refer to the table in Figure 10-3. For a solenoid to provide linear motion, it must receive either an AC or a DC pulse. Where AC is used, the time constant of the solenoid's action must be long enough to prevent chattering. Some DC solenoids will rattle when 60 Hz is applied to them.

The pneumatic and the hydraulic actuators listed in Figure 10-5 are essentially air or fluid valves that are controlled by a solenoid. The output of the electronic control function simply operates the solenoid. The resulting translation of the valve opening or closing into a hydraulic or pneumatic energy change actually has nothing to do with the electronic troubleshooting portion of industrial equipment. If proper voltage is applied to the solenoid and the valve does not work, the entire actuator must usually be replaced.

10.3 HOW TO USE THE SYMPTOM-FUNCTION TECHNIQUE

Now that you know how to deal with the input and output signals of industrial controls and instruments, you can learn to use the symptom-function technique. In previous chapters we have explained this technique in detail and have shown you how to use it for troubleshooting transistor circuits, integrated circuits, and electron tube circuits. The vast majority of industrial controls and instruments already uses transistors, a few equipments still use vacuum tubes, and some of the latest equipments also incorporate ICs. You don't even need an instruction book to find out what makes up the active portions of the control functions of a particular equipment. Just look at the open chassis.

The symptom-function technique is the most powerful tool we have to isolate troubles in industrial controls and instruments, but we must be sure to properly determine which are the symptoms and which are the functions. In troubleshooting TV receivers, the symptoms were observed on the screen or heard on the loudspeaker. In digital equipment, the symptoms could be determined by the results. Unfortunately, where feedback circuits are concerned, it is more difficult to determine which is the symptom and which the disease. If you expect to troubleshoot servo systems to any great extent, we recommend the book *Servomechanism Fundamentals and Experiments*, Philco Tech. Center, Prentice-Hall, Inc., 1964.

Troubleshooting servo systems is a very specialized field. The following examples illustrate how the symptom-function technique can be applied to industrial controls.

The first example is illustrated in Figure 10-6 and represents the mixing vat in a food-processing plant. To simplify the problem, we have assumed that only two liquids are being mixed. Each of the liquids comes from a different storage tank and is pumped through different lengths and different diameter pipes to the mixing vat. The flow of the liquid through the pipe is controlled, in each case, by a motor-driven valve. If it is desired that the same flow, gallons per minute, be supplied through both pipes, the output of flow meter 1 must equal the output of flow meter 2. A solid-state signal comparer and control section compares the two voltages, corresponding to the flow of the liquid through the two pipes. If the voltage from flow meter 1 becomes greater, the servo motor drive amplifier connected to the motor-driven valve No. 1 will activate the motor to turn the valve down. Similarly, if flow meter No. 2 shows excessive output, the valve controlled by motor No. 2 will be shut down. Specific minimum and maximum flow meter signal-level settings are provided at the signal comparer. Without such limits, an overshoot in the servo

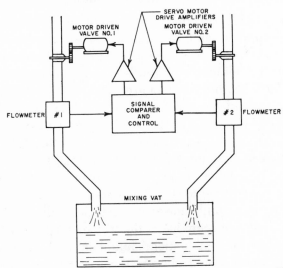

Electronics World

FIGURE 10-6
ELECTRONIC CONTROL FOR A MIXING VAT

amplifier could cause motor-driven valve No. 1 to shut down too much, reducing the flow in pipe No. 1. When this is compared to flow meter No. 2, its servo amplifier would shut down valve No. 2 some more. Within a short time both valves could be shut down completely. The control portion of the electronics contains reference voltages, similar to the speed-control input to the differential amplifier in Figure 10-2, to prevent shutdown or excessive valve opening.

We have received the customer's complaint that the motor-driven valve No. 1 has the tendency to shut off the flow in that pipe after the equipment has been operating for a few hours. The motor-driven valve on No. 2 appears to operate properly. It becomes immediately clear that the difficulty is in the circuit that controls the flow through pipe No. 1. Because of this defect both pipes are shut off at the remote end. We cannot check the output of either flow meter since nothing flows through the pipe. Applying symptom-function reasoning, we can eliminate defects in the circuitry controlling No. 2 pipes. Remember that only the No. 1 motor-driven valve tends to shut down. Remember also that this defect only appears after the equipment has been in operation for a while. Experienced electronics troubleshooters will immediately suspect a temperature problem, a defect that usually is only apparent after a sufficient warm-up period.

A preliminary visual inspection of the circuitry in the signal comparer and control equipment and, in particular, the servo motor drive amplifier, might reveal an overheated resistor or some other clue. By deduction, we can eliminate either flow meter 1 or flow meter 2 as a source of defect. Neither of them are going to get warm and even if they did, this should not cause the motor-driven valve No. 1 to be shut down altogether. Using the same logic, we can see that that portion of the signal comparer that handles both the No. 1 and No. 2 flow-meter signals must also be ok. Any defect in that circuitry would be likely to cause a defect in both sides. If the comparer itself were off-balance, it would tend to shut down one valve and open the other fully. We have been told, however, that motor-driven valve No. 2 works fine. It does not seem likely that the motor-driven valve itself, an electromechanical combination, should fail in this manner. In any event, its positioning is controlled by the servo motor drive amplifier. The most likely suspect, without performing any detailed tests, is therefore the servo motor drive amplifier going to valve No. 1.

Figure 10-7 shows, in simplified schematic form, a typical thickness

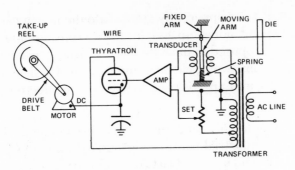

FIGURE 10-7
WIRE THICKNESS CONTROL SYSTEM

control system for wire-pulling machinery. The wire is pulled through a die by the take-up reel, which is driven by a motor. The torque of this motor is controlled by a DC voltage, which is derived through a thyratron rectifier from the main AC power. The thickness gauge for measuring the wire, as it emerges from the die, is a differential trans- former transducer, energized by the same 60 Hz AC power as is applied to the thyratron plate. In this case, the amplitude of the AC output of the transducer is linearly related to the wire thickness. The transducer drives a small, solid-state amplifier, which, in turn, pro- vides the firing voltage at the grid of the thyratron. If the wire is too thick, the moving arm of the transducer transformer lowers an iron core into the transformer and this increases the control voltage ap- plied to the amplifier. In turn, this increases the grid-control voltage of the thyratron and therefore the amount of DC voltage applied to the motor. With more DC input, the motor can turn the reel faster and this tends to thin out the wire. The reference signal going into the control amplifier is set for the desired wire thickness.

The customer complained that he could vary the setting of the refer- ence signal and get corresponding speed changes. Left on its own, however, the system did not control the thickness of the wire. Using the symptom-function technique, we deduce that the thyratron and the control amplifier must be operating correctly, since the reference setting changed motor speed. Either the transducer was defective or else the transducer was not getting its excitation signal from the power transformer. We know that mechanical defects are more likely than electronic defects and for that reason we first look at the fixed and moving arm of the transducer itself. Jiggling the moving arm with a screwdriver indicated that it did not move freely. In fact, an accumulation of dirt had apparently gotten into the moving arm shaft so that it could only be moved up and down with considerable force. The spring that was supposed to push the moving arm up may have lost its strength, too. In any event, cleaning the moving arm shaft and replacing the spring cleared up this trouble without any electronics work at all.

These two examples of using the symptom-function technique for troubleshooting industrial controls and instruments illustrate that in this field there are no hard and fast rules, but that you must figure out the symptoms and their related functions, based on the block diagram and the operation of the particular system.

10.4 LIMITATIONS OF SIGNAL TRACING

If we look at the block diagrams of typical industrial control systems, such as in Figures 10-2, 10-6, or 10-7, we suspect that the signal-tracing technique will only be of limited use. Taking the last illustration first, we can, of course, use a volt-ohmmeter to check that the transformer has the proper output voltages at each winding and tap. The input to the amplifier, however, is at a very low level and it is difficult to relate the exact voltages to wire thickness. We would have to use a precision spacing gauge in place of the wire to determine if the input signals to the differential amplifier are analogous to wire thickness. To check the operation of the thyratron we would need an oscilloscope in order to measure the AC voltage on the control grid and on the plate and then compare the resultant motor control voltage with the thyratron manufacturer's data. We would need a tachometer to measure the motor speed and relate that to the control voltage, based on the characteristic curves of the motor. Clearly this type of signal tracing is slow and cumbersome.

The electronic control unit for the mixing vat system in Figure 10-6 may allow us to use the signal-tracing method if we had a detailed circuit diagram for it. Even then, however, we would need to simulate the outputs of both flow meters in order to trace the various control voltages through. The same thing applies to the basic industrial control circuit shown in the block diagram in Figure 10-1.

Signal tracing has its limitations in industrial control systems, partly because the signals are relatively simple and partly because of the feedback element that exists in most such systems. The technique recommended for troubleshooting any kind of servo system where a closed loop or feedback exists is to open the loop and measure signal levels along the servo path. If the measurements correspond to the open loop values given in the manufacturer's manual, we can assume that these portions are operating correctly.

When you troubleshoot so-called "open-loop" control systems, the signal-tracing method has more merit. Even in these systems, however, the signals that you will trace through are relatively simple and do not change as much in form or frequency as those in TV and hi-fi equipment. Many open-loop systems are of the digital type when counting or timing are the essential functions to be performed. When you run into these systems, deal with them as you did for digital equipment, for which the troubleshooting procedures are described in chapter 9.

10.5 HOW TO USE THE VOLTAGE-RESISTANCE TECHNIQUE

Next to the symptom-function technique this approach is the most powerful and widely used one in troubleshooting industrial controls and instruments. Unless you are familiar with the particular piece of equipment, however, you should not even try to make voltage and resistance measurements without having the detailed manufacturer's data on hand. Industrial control systems usually deal with widely varying power levels, such as the power used to drive the take-up reel in Figure 10-7, as compared to the control-signal power derived from the thickness gauge itself. This means that the voltage measurements may have to be performed with different instruments. The physical load conditions also play a great part. Before starting a set of measurements, look carefully in the manufacturer's data for the conditions under which these measurements are supposed to be taken. In some instances two or three sets of voltage values are given, one for full-load conditions, one for half-load, and the third for no-load conditions. Make sure that the equipment loading conditions are exactly as specified before you try to compare the voltages measured with those stated in the manufacturer's data. In many instances, actual voltage values may be less important than the variations of voltage with varying load.

Another important consideration in voltage and resistance measurements is impedance of the test equipment. If this is specified in the manufacturer's data, be sure that your meter complies. Most transducers have a relatively low impedance, but where high impedance outputs are used the wrong meter may load down the circuit. Watch out for tranducers that use a bridge arrangement, a common practice among strain gauges in particular. If you connect a low-impedance meter across one-half of the bridge, this will unbalance the entire system and wrong readings will result.

Strain gauges are so widely used in industrial equipment that a few words of caution are in order here. Figure 10-8 shows a typical strain gauge bridge. You may be tempted to test this bridge simply by measuring the DC resistance of each leg and you may find that each of them measures 100 ohms. Under full load, however, one side of the bridge may be unbalanced by a resistance change of only 2 to 3 ohms. This may be sufficient for the bridge to operate properly, but an ordinary ohmmeter will hardly indicate such a small difference. If we use the standard excitation signal, usually an AC voltage, and measure the differential amplifier output under no-load and full-load conditions, the difference in output signal will show whether the bridge works correctly or not. The ohmmeter measurement merely establishes continuity.

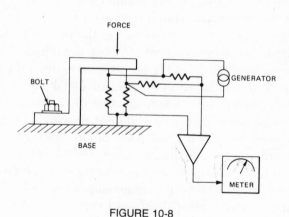

FIGURE 10-8
STRAIN GAUGE BRIDGE

Another widely used building block in industrial control systems is the voltage and the power amplifier. In essence, these are usually the same kind of AC amplifier that you find in audio and hi-fi equipment. When you are trying to troubleshoot an amplifier circuit, refer back to chapter 8 for the techniques that work best with these amplifiers.

10.6 HOW TO FIND THE DEFECTIVE COMPONENT

Once you have isolated the defect to a particular portion of the equipment, you want to locate the defective component in the most efficient way. The basic substitution technique has been discussed in chapter 1 and its application to transistors and integrated circuits has been described in more detail in chapters 3 and 4 respectively. Of course, if a vacuum tube equipment is used, replacing tubes, one-by-one, is the most obvious procedure to follow. If you refer back to chapter 5, however, you will see that it is not always possible to properly test all tubes. In the case of thyratrons and some of the other large industrial electron tubes, the tube tester may be useless. Some of these tubes are quite expensive and if you simply substitute a new one, a circuit defect may ruin that tube, too. Before running such a risk, you should first measure the voltages, at least on the control elements of the tube. In some instances it is possible, by visual inspection, to determine if some internal element of the tube is broken or loose. Remember that many industrial tubes are filled with gas and are expected to give off a glow. In a TV set a bluish glow in a power rectifier indicates a defect, but in a thyratron this may be its normal operation.

In many industrial control systems the electronic circuitry is contained on plug-in P.C. modules and it is then possible to replace an entire module. Because equipment downtime can be so costly, most plants that use electronic controls also keep a supply of spare parts, including P.C. modules, on hand. These plug-in spares are very useful for troubleshooting. Relays, in particular, are often used with plug-in sockets and, whenever you suspect a relay defect, you will want to plug in a spare relay.

If you are reasonably sure that none of the electronic parts, such as resistors, capacitors, transistors, etc., are defective, connectors and cables are the next likely suspects. The best way to check them is with an ohmmeter. Be sure to test for continuity as well as for shorts between pins, defective insulation, cold solder joints, and similar mechanical defects.

We have found that transformers of all types are more frequently defective in industrial equipment than they are in some entertainment or digital equipment. Be sure to look for loose laminations, broken wires, and charred insulation in or around the transformers.

10.7 MOST FREQUENTLY FOUND TROUBLES IN INDUSTRIAL CONTROLS AND INSTRUMENTS

It is difficult to provide a list of most frequently found troubles in industrial equipment because of the wide variety of applications. In electronic weighing systems, for example, the load cells or strain gauges themselves are usually the most frequent sources of the defect. In chemical plants, particularly where corrosive chemicals are used, the most frequent defects seem to be due to corrosion of electronic components, connections, grounds, etc. In electronic heat-sealing or induction-heating equipment, RF power tubes are used to generate the required energy. These tubes have a limited lifetime and are therefore the most frequent trouble source. Mechanical transducers of all types are more failure-prone than photoelectric transducers, with the exception of the incandescent lamps used in connection with photocells. Temperature transducers fail relatively rarely. Among the actuators, solenoids fail more often than motors, and both pneumatic and hydraulic actuators frequently fail in their valve but not in their solenoid portions. Whenever relays are used they are very likely trouble sources, even though relays, by themselves, are extremely reliable devices. Telephone-type relays, for example, are often rated for a million operations or more. Industrial relays, however, possibly because they control larger currents, seem to fail more often.

Experienced troubleshooters agree that in industrial electronic equipment mechanical failures occur much more frequently than electronic failures. The reasons for this are mechanical wear, friction, metal fatigue, abrasion, and similar effects. Look for moving parts and make sure that they move properly. Look for springs that are broken or have lost their tension. Look for worn shafts, frayed or worn drivebelts, loose bearings, loose pulleys or gears, or similar mechanical problems. In general, linear motion is harder on mechanical parts than rotary motion.

When you see excessive vibration, mechanical noise, metal filings, or dust around the control elements or the actuators of the defective system, inspect all of the mechanical portions very carefully. Often a spring has broken and snapped off and, unless you know that there should have been a spring, you may not find this trouble. If the bearings on a tachometer are worn or if a shaft is slightly bent, its electrical output signals may be incorrect.

Occasionally we have found that the plant maintenance personnel, in trying to repair a particular piece of equipment, has modified it beyond recognition. Be sure to point this out to the customer and explain why the repair will now be more costly. We have found wrong tubes plugged into sockets, capacitors replaced with wrong value components, and similar signs of improper repair, even though the customer has assured us that no one has previously tried to repair this equipment.

10.8   A LAST-RESORT METHOD FOR TROUBLESHOOTING INDUS-TRIAL CONTROLS AND INSTRUMENTS

Suppose you have worked for hours trying to locate the trouble in some industrial equipment. You have tried all of the different techniques and have still not located the source of the defect. Now the time has come to look at what you have done and what you have forgotten. Write down, to the best of your recollection, all of the parts that you have replaced, all of the changes you have made and all of the measurements you have taken. This is a good time to take a break and possibly continue the troubleshooting effort the next day. This gives our subconscious a time to rehash our thinking and possibly come up with the mistake that we must have made.

Before you start troubleshooting again, go over the manufacturer's manual with great care, look over the original block diagram, review each of the functions of the equipment, and see how the manual corresponds to the actual equipment. We recall one particularly nasty job in which we finally realized that the manufacturer's manual really described a much later model of the same equipment. A stage of feedback control had been added and our open-loop gain measurements were therefore completely different from the values given in the manual. Eventually the defect turned out to be excessive friction on an idler in the gear train.

Consider the possibility that one of the replacement parts or replacement modules has been defective itself, adding complexity to the search for the real defect. Remove these replacement parts and install the original components again, one by one. After each such step, check out the system again to see if the original defect is there or if a change has occurred.

With the correct manufacturer's manual at hand, make a visual inspection of every circuit element is every part of the equipment. Next, check to see if your test equipment meets the specifications indicated by the manufacturer. With the correct test equipment, see if you can duplicate each of the voltage and current measurements indicated in the manufacturer's manual. When you review these figures you may find, for example, that all your readings are high, low, or off in some other way. This may indicate a marginal power supply.

Some industrial equipment can be operated under varying load conditions and, in some cases, under special test-load conditions. If it is possible, try to set up the test-load conditions and check the results against those stated in the manufacturer's manual. If the equipment operates properly under test-load conditions, make sure that the full-load conditions under which the equipment normally operates are within those specified by the manufacturer. In the case of the wire-pulling system, illustrated in Figure 10-7, the die size may have been changed and the wire that is now being pulled is heavier than was originally specified. The space between the fixed and the moving arm of the transducer may readily accept the heavier wire. This, however, may place the movable iron core so far down into the differential transformer that slight variations in wire thickness do not have enough of an effect on the differential transformer.

Even though you have checked all of the mechanical linkages of the system before, check them again as thoroughly as possible. A rotating shaft may move freely when there is no load on it or at low speeds, but may encounter considerable friction when load is applied to it or when it is rotated at high speeds. Remember, mechanical trouble is more likely than electronic trouble.

Check the power supply voltage when the entire plant is working, as well as during off-hours. Sometimes the difference in AC line voltage is so great that the internal power supply regulating circuit drops out of regulation at extremely low line voltage. For the 117 volts AC supplies, 105 volts is usually the lower limit at which the equipment can be expected to work. Even though the equipment can work at this level, its performance may be marginal and accuracy of control may be lost because critical reference voltages are out of calibration.

Because of the great diversity among industrial control equipment, we cannot give you a single last-resort technique that will apply to every kind of equipment. If you suspect that the trouble is due to any kind of interference, review chapter 7 and see how it can apply to the problem at hand. Try to find out if any new electronic equipment, particularly of the type that would cause interference, has recently been installed. Remember that the interference can enter either by means of the AC power line or by radiation. If you suspect that the defect is intermittent, refer to chapter 6, which contains directions on how to find this type of defect.

Chances are that several equipments of the type that you are troubleshooting are located in the plant. It is often possible to interchange portions of the equipment, such as modules, subassemblies, control portions, etc. Customers do not like it when we have to disable another control equipment in order to troubleshoot the one already down. As a last-resort method, however, this approach often identifies at least the particular portion of the equipment in which the defect occurs. Frequently, comparison measurements between the two equipments permit you to isolate the defect.

Review the entire circuit diagram and try to isolate those components that could cause a number of other components to fail as well. You can also usually identify those critical components where even a small change in value can affect the overall performance. It may be necessary to remove such components, such as precision resistors and capacitors, and to measure them carefully to make sure that they still have the correct value. The feedback network in a servo amplifier, the input resistors in a differential amplifier, and the frequency determining elements in an oscillator are typical of this type of critical component. Consider the possibility of a build-up of tolerances in several stages of the circuit that could cause a malfunction.

No matter how difficult a troubleshooting job may seem, remember that the equipment did work correctly once and it must therefore be possible to repair it. If a man made it work once, you can make it work again.

Index

INDEX

226